The Biodiversity **Crisis**

The Biodiversity **Crisis**
Losing What Counts

Michael J. Novacek, Editor

AN AMERICAN MUSEUM ö NATURAL HISTORY BOOK

The New Press, New York

Published in the United States by The New Press, New York, 2001
Distributed by W. W. Norton & Company, Inc., New York

ISBN 1-56584-570-6 (pbk.)
CIP data available.

The New Press was established in 1990 as a not-for-profit alternative
to the large, commercial publishing houses currently dominating
the book publishing industry. The New Press operates in the public
interest rather than for private gain, and is committed to publishing,
in innovative ways, works of educational, cultural, and community
value that are often deemed insufficiently profitable.

The New Press, New York
450 West 41st Street, 6th floor
New York, NY 10036

www.thenewpress.com

Printed in England

9 8 7 6 5 4 3 2 1

Contents

Section One: What Have We Lost,
                  What Are We Losing?

Section Two: Extinctions Past and Present

Section Three: Saving Biodiversity: Strategies and Solutions

Foreword Ellen V. Futter

Since its founding in 1869, the American Museum of Natural History has put the world on display. The twin pillars of our mission have always been science and education. Today, the Museum is one of the world's leading research institutions in the natural sciences. Over the years, our scientists—now a staff of more than 200 men and women—have gone on more than 100 expeditions a year. They collect evidence from all over the globe in their effort to answer questions about such fundamental scientific and human issues as the origins of the universe, Earth and life, who we are and where we fit. In addition to research, our scientists have a related responsibility—to interpret science for the general public. The exhibitions at the Museum have been conceived and curated by scientists who are committed to putting the evidence— "the real stuff"—in front of the public.

As we move into the twenty-first century, we at the Museum are filled with a renewed dedication to our mission. To many people, science today seems too removed and too difficult to understand, yet the need has never been greater for a public that is well informed about science. Through its educational initiatives and exhibitions, the Museum seeks to narrow the gap between what people know and what they need to know about science. To that end, in 1997 the Museum launched the National Center for Science Literacy, Education, and Technology to extend the Museum's resources beyond its walls to a national audience.

Science is exploration. Scientists work at the frontier—at the border of the known and the unknown. This book series, through the words of working scientists, enables non-scientists to share the excitement of cutting-edge science, the excitement of discovery. The series includes four volumes that expand the themes covered in many of our major new exhibitions. Our exhibitions always embody a scientist's vision and point of view. In the same way, each book in this series is "curated"—researched, organized, and introduced—by one of the Museum's scientists. Each book features a selection of essays written by leading scientists who have made significant contributions to the field. The essays are supported by case studies and profiles of important people and events.

This volume, *The Biodiversity Crisis: Losing What Counts*, addresses the main themes of the Museum's permanent Hall of Biodiversity, which opened in May 1998 to great critical acclaim. The book, like the hall, focuses on the variety and interrelatedness of all living things on our planet. Other volumes in this series include: *Epidemic!: The World of Infectious Disease*, which explores the themes of a major temporary exhibition by the same name at the Museum in 1999; *Earth: Inside and Out*, which focuses on the Earth and the phenomena that shape our planet to expand the themes addressed in the Gottesman Hall of Planet Earth; and *Cosmic Horizons: Astronomy at the Cutting Edge*, which explores the mysteries and wonders of the universe to expand the themes addressed in our new Cullman Hall of the Universe.

This series illustrates our continuing commitment to connect the general public with the natural world. We cannot send real specimens to every home and classroom, but we can bring the ideas, concerns, and questions of working scientists directly to you. We hope these books provide a valuable resource that will prepare tomorrow's leaders to make informed decisions about the world we all share.

Ellen V. Futter is President of the American Museum of Natural History.

Acknowledgments

This book was made possible by a generous grant from the Garden Club of America. The original volume, *Scientists on Biodiversity*, was developed by the National Center for Science Literacy, Education, and Technology and the Center for Biodiversity and Conservation, both at the American Museum of Natural History, in conjunction with the opening of the Hall of Biodiversity, a permanent exhibition at the Museum. The Garden Club's support enabled the Museum to create this resource on biodiversity science for a broad audience and to distribute the book widely by many Garden Club of America members and through the Museum's teacher development activities. We especially acknowledge the work of Francesco Grifo and Eleanor Sterling for their contributions to the original edition.

This new version represents a major reorganization of the original contributions into three main sections, with an introduction to each by Michael J. Novacek. Also included are new formats for illustrations, topic outlines, study questions, and information concerning the research, exhibitions, and educational programs at the American Museum of Natural History. In this way, the new volume is intended to reach an even broader audience of high school students, teachers, and the general public.

This book was produced by the National Center for Science Literacy, Education, and Technology, American Museum of Natural History.

Ellen V. Futter, President

Myles Gordon, Vice-President for Education

Nancy Hechinger, Director of the National Center for Science Literacy, Education, and Technology.

Production Staff:
Project Director: Caroline Nobel
Creative Director: Patricia Abt
Production Manager: Ellen Przybyla
Production Coordinator: Eric Hamilton
Production Assistant: Ethan Davidson

Original Editorial Staff:
Content Developers: Francesca T. Grifo and Eleanor J. Sterling
Editors: Linda Koebner and Jane E. S. Sokolow
Senior Producer and Editor: Sharon Simpson
Science Editor: Kefyn M. Catley
Writer: Theron Cole, Jr.
Education Editor: Tim O'Halloran
Assistant Editor: Phillip Fujiyoshi

Design by Sheena Calvert, parlour design, NY

We would like to thank Diane Wachtell, Ellen Reeves, Barbara Chuang, Gary Tooth, Leda Scheintaub, and Fran Forte at The New Press.

The National Center would like to acknowledge the National Aeronautics and Space Administration for its general programmatic support and for support for this book series.

Exterior of the American Museum of Natural History.

"Biological diversity is the key to the maintenance of the world as we know it." — Edward O. Wilson

Tree frog (*Litoria infrafrenata*)

Preface: Biodiversity Michael J. Novacek

When the last millennium began, the world was mostly uncharted: mysterious, even terrifying in its vastness outside clusters of human habitation. Nature over large portions of continents and oceans was wild and untamed, but at the same time protected from the invasion of a comparatively concentrated human population. At the end of the millennium, of course, the tables are turned, and humans, through their sheer numbers and their capacity for altering the natural state of the planet, present a threat to the biosphere that rivals some of the great extinction events of the past.

What exactly is at stake? Earth is home to tens of millions of living species, of which we are but one—a legacy of 3.5 billion years of evolution. Biodiversity is the spectacular variety of life on Earth and the essential interdependence among all living things. First formally used at a 1986 forum of researchers sponsored by the National Academy of Sciences in the United States, it has become the most commonly used word of scientists, conservationists, educators, and policy makers to describe a scientific discipline and an approach, as well as a critical—indeed, life-threatening—issue. We know that because of human action and intervention we are losing biodiversity at an alarming rate. In a recent American Museum of Natural History survey conducted by Louis Harris and Associates, seven out of ten life scientists rate biodiversity loss as the single most important issue for the twenty-first century.

Scientists say we are in the middle of a sixth major mass extinction. (The last great extinction event occurred at the end of the Cretaceous Period, about 65 million years ago, when an estimated two-thirds of all species, including all the dinosaur groups except the birds, were obliterated.) The same survey shows that the general public has little awareness of the magnitude of the loss, or its implications for life on Earth.

Indeed, the dimensions of biodiversity loss and its implications are not well understood. The biological world is largely unknown. Most of the larger animal and plant species have been thoroughly described, but scientists know significantly less about the insects, spiders, soil organisms, bacteria, and other microbes that play a critical role in the functioning of ecosystems. It is estimated that less than thirty percent of the millions of species that live in the different habitats of the United States have been discovered and described. Worldwide, the total is estimated to be less than fifteen percent.

Scientists may disagree on the numbers—how many species there are, how many are being lost—but even conservative estimates predict that global species loss could reach as high as thirty percent of total species by the early decades of the next century, by the time today's children have children. A loss on such a scale is certain to have serious consequences for the recycling of nutrients on the planet, for the production of new food sources and medicine, for the quality of our air, water, and soil, and for the overall quality and viability of life for humans and other species. As many of the articles in this book point out, human action is at the root of the biodiversity crisis. Yet perhaps our human talent for problem solving can provide the solution. We have the capacity for stewardship of the planet, and we are empowered to improve—or at the very least sustain—the balance of life on Earth if we have the knowledge and the will.

Michael J. Novacek is Senior Vice President and Provost of Science and a Curator of Paleontology at the American Museum of Natural History.

What we don't know can hurt us. We need to know more about the biological world. At the American Museum of Natural History, we are dedicated to researching the many unexplored dimensions of biodiversity and bringing our findings to the general public, as well as to the scientific community. Of equal importance is our commitment to convincing the widest possible audience that the protection of biodiversity is the foremost scientific challenge as we enter the twenty-first century. Many of our 200 scientists are devoted to exploring, discovering, and describing the biological world in the branch of science called systematics. These women and men are working within a tradition that extends back to the founding of the Museum in 1869.

The structure of *The Biodiversity Crisis: Losing What Counts* mirrors the structure of the Museum's Hall of Biodiversity. "What Have We Lost, What Are We Losing?" defines biodiversity and its importance to the balance of the planet and the welfare of our species. "Extinctions Past and Present" describes the present crisis in the context of the five former mass extinction episodes in Earth's history, and describes in detail some of the major threats to biodiversity. The final section, "Saving Biodiversity: Strategies and Solutions," stresses the positive steps being taken by scientists and others to stem the loss of biodiversity.

One part of stewardship is passing on knowledge—maybe even a little wisdom—from one generation to the next. The articles in this book were written by many of today's leading scientists in the field of biodiversity. They seized this opportunity to speak directly in their own voices with their own diverse perspectives to today's youth—tomorrow's stewards.

Above: Spotted frog (*Rana pretiosa*)

Below: Eastern spadefoot toad (*Scaphiopus holbrooki*)

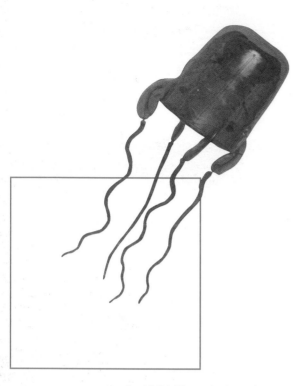

Box jellyfish *(Tamoya haplonema)*

Section One: What Have We Lost, What Are We Losing?

Protea (*Protea* sp.), South Africa

Introduction Michael J. Novacek

When it comes to biodiversity, ignorance is not bliss. In this section, while we acknowledge the depth of our ignorance about life on this planet, enough is known to indicate life is in crisis. The authors cast the ongoing loss of habitats and species in dire terms, on a scale comparable to the mass extinctions of the past. It is as normal for species to become extinct as it is for new species to evolve. Both the rates of evolution and extinction of species are relatively low. In periods of mass extinction, it is the increased rate of extinction that is significant. For example, Norman Myers claims that the current rate of species loss is probably 10,000 times greater than that of the background extinction rate, or the normal extinction rate. It is also possible to identify in many dimensions just what we are losing—namely, the enormous benefits of biological diversity in terms of eco-system services, products for human health and consumption, and enhancement of the quality of life in both economic and aesthetic terms. As Edward O. Wilson notes in the opening essay of this section, wild species are the source of forty percent of our medicines, and all species offer unique insights to science. Indeed, as Francesca T. Grifo emphasizes, biodiversity is not only a priceless source of pharmaceuticals, it is critical to maintaining safe water supplies and controlling disease-causing organisms.

The integral importance of millions of species in maintaining processes essential for life— control of toxins and harmful microbes, soil decomposition, and recycling of nutrients and atmospheric gases—is a recurrent theme in essays by Norman Myers, Paul R. Ehrlich and Simon A. Levin, and Peter H. Raven. It should be remembered, however, that the benefit of a given ecosystem is not simply a function of how many species live in a particular region. David Ehrenfeld vividly describes the unspoiled Soper Valley in Baffin Island, Canada, which remains an environment with its "...own kind of grandeur and glory, a living presence that surely makes an equally powerful claim on our care and affection," despite the low number of species.

Ignoring, or irresponsibly exploiting, these benefits—medicinal, economic, or aesthetic— has drastic effects. Robert Repetto explains why unsustainable logging practices not only create unnecessary waste of trees and other forest products, but also increase greenhouse gases through burning of huge tracts of tropical forests. Thus, loss of species does not only translate to loss of valuable natural habitat; as Peter Raven notes, this crisis threatens the basic physical system of the planet, involving the recycling of chemicals in the soil, oceans, and atmosphere.

As all the contributors stress, one way to mitigate the massive degradation and destruction of habitats is to infuse conservation action with fundamental scientific discovery. Thomas Eisner describes developments in molecular biology and biochemistry that help scientists identify chemicals that are used by species in defense against predators and other adaptations. Chemical substances once discovered can be used in their natural state or can provide a model for synthesized compounds. The current range of such chemical applications is impressive and the potentials for chemical exploration of the biosphere are extraordinary. Such work puts a tangible value on biodiversity in terms of human need, and moves us from short-term, indiscriminate harvesting of habitats toward bio-exploration and sustainable harvesting of new products. Science must set the framework for such sustainable programs. For example, comprehensive scientific data on population sizes and viability in species of ocean fishes and other marine life can identify both good and bad management, as effectively shown in the

Audubon Guide to Seafood compiled by Carl Safina. The contributors to this section as well as the scientists, artists, and educators profiled have dedicated their careers to the advocacy of sound, scientifically based conservation of biodiversity. It is hoped that their words will inspire future generations of biodiversity scientists as well as policy makers.

To explore what we have lost and what we are losing, I pose the following questions:

Why is biodiversity in crisis?

Edward O. Wilson, Pellegrino University Research Professor at Harvard University, describes the causal factors behind today's biodiversity crisis and suggests possible solutions.

What does biodiversity do for us?

Norman Myers, an independent scientist and a consultant who has worked on questions of the environment and development for more than twenty years, explains how species diversity benefits our health, our food production, and even our economies.

How does random and unregulated deforestation in the tropics affect our lives?

Robert C. Repetto, visiting Professor in the School of Forestry and Environmental Studies at Yale University, demonstrates that bad logging practices not only create unnecessary wastage of trees and other forest products, but also contribute to the greenhouse effect.

How does human health benefit from a wealth of biodiversity?

Francesca T. Grifo, who holds an adjunct faculty appointment at Columbia University, explains the importance of biodiversity to our food and water supplies in helping us to prevent and cure disease.

Why is biodiversity crucial to our well-being?

Paul R. Ehrlich, Bing Professor of Population Studies in the Department of Biological Sciences at Stanford University, and **Simon A. Levin**, George M. Moffett Professor of Biology and the Founding Director of the Princeton Environmental Institute at Princeton University, describe the complexity and importance of species ecosystems and ecosystem services.

Is it necessary to always emphasize species diversity as the main reason for conserving environments?

David Ehrenfeld, Professor of Biology at Cook College, Rutgers University, suggests that an environment like the Soper River Valley in Baffin Island, Canada, may be poor in species diversity but rich in natural beauty.

Why should we care about biodiversity loss?

Peter H. Raven, Director of the Missouri Botanical Garden and Engelmann Professor of Botany at Washington University, argues that it is our social responsibility to be aware of the destruction we are causing to the environment, and that we need to become custodians of Earth and of all its inhabitants.

What can we learn from the natural world?

Thomas Eisner, Jacob Gould Schurman Professor of Chemical Ecology at Cornell University, emphasizes the need for people to become explorers of the chemical properties of substances found in the natural world.

Acantharian (lynchnaspis gilltschi)

Biodiversity: Wildlife in Trouble

Edward O. Wilson

Around the world, biodiversity—defined as the full variety of life, from genes to species to ecosystems—is in trouble. Not a week goes by without reports of the imminent end of one species or another. For every celebrity animal that vanishes, biologists can point to thousands of species of plants and smaller animals either recently extinct or on the brink.

There is not one country, not one biome and its plant and animal life—mountain, desert, or ocean—that remains untouched. The rarest bird in the world, the Spix's macaw, is down to one

Edward O. Wilson is Pellegrino University Research Professor and Honorary Curator in Entomology at Harvard University.

or possibly two individuals in the palm and river-edge forests of central Brazil. The rarest plant is Cooke's kokio of Hawaii, a small tree with orange-red flowers that once lived on the dry volcanic slopes of Molokai. Today it exists only as a few half plants—branches grafted onto the stocks of other related plants. Despite the best efforts of scientists to save the plant, no branches planted in soil have sprouted roots.

It is difficult to estimate overall rates of extinction. However, biologists generally agree that on the land, at least, and on a worldwide basis, species are vanishing one hundred times faster than before the arrival of humans. The world's flora and fauna are paying the price of humanity's population growth.

Biodiversity is in serious trouble. Responding to the problem, conservation experts have shifted their focus in the past twenty years from individual plant and animal groups (species) to entire threatened habitats, whose destruction would cause the extinction of many species. Such "hot spots" have become the focus of conservation efforts. The logic of the experts is simple: by concentrating conservation efforts on such areas, we can save the largest amount of biodiversity at the lowest economic cost. The outright elimination of habitats is the leading cause of extinction. But the introduction of exotic species and the diseases they carry follows close behind in destructiveness, along with overhunting or overharvesting of plants and animals. All these factors work together in a complex manner. When asked which ones caused the extinction of any particular species, biologists are likely to give the *Murder on the Orient Express* answer: They all did it. A common sequence in tropical countries starts with the building of roads into wilderness. Land-seeking settlers pour in, clear the rain forest on both sides of the road, pollute the streams, introduce alien plants and animals, and hunt

wildlife for extra food. Many native species become rare, and some disappear entirely.

People commonly respond to the evidence of species extinction by entering three stages of denial. The first is, simply: Why worry? Extinction is natural. Species have been dying out for billions of years without permanent damage. Evolution has always replaced extinct species with new ones.

These statements are true—but with a terrible twist. After each of the four greatest environmental disruptions that occurred during the past 400 million years, evolution needed about ten million years to restore Earth's biodiversity. Worse, evolution will be even slower if natural environments have been crowded out by artificial ones. Faced with such a long waiting time, aware that we have inflicted so much damage in a single lifetime, our descendants are going to be—how best to say it?—peeved with us.

Entering the second stage of denial, people ask: why do we need so many species anyway? Why care, especially since the vast majority are bugs, weeds, and fungi?

It is easy to dismiss the creepy crawlies of the world. However, the value of the little things in the natural world has become extremely clear. Recent experimental studies on whole ecosystems support what ecologists have long suspected: The more species living in an ecosystem, the higher its productivity and the greater its ability to withstand drought and other kinds of environmental strain. Because we depend on working ecosystems to cleanse our water, enrich our soil, and create the very air we breathe, biodiversity is clearly not something to discard carelessly.

Besides creating a livable environment, wild species are the source of products that help

Calcareous alga (Halimeda incrassata)

support our lives. More than forty percent of all prescription medicines used by Americans are substances originally extracted from plants, animals, fungi, and microorganisms. Aspirin, for example, the most widely used medicine in the world, was originally derived from the willow tree.

Every species on Earth is a masterpiece of evolution, offering a vast source of useful scientific knowledge because it is so thoroughly adapted to the environment in which it lives.

Even when that much is granted, the third stage of denial usually emerges: Why rush to save all the species right now? We have more important things to do. Why not keep live specimens in zoos and botanical gardens and return them to the wild later?

The grim truth is that all the zoos in the world today can hold a maximum of only 2,000 species of mammals, birds, reptiles, and amphibians, out of about 24,000 known to exist. The world's botanical gardens would be even more overwhelmed by the quarter million plant species. To add to the difficulty, no one has come up with a plan to save all the insects, fungi, and other ecologically vital small organisms. And even when scientists are finally ready to return species to independence, the ecosystems in which many live will no longer exist. The conclusion of scientists and conservationists is practically unanimous: the only way to save wild species is to maintain them in their natural habitats. Considering how rapidly such habitats are shrinking, even that straightforward solution will be an overwhelming task.

In spite of all these difficulties, there is reason for some optimism. With appropriate measures and the will to use them, the destruction can be slowed, perhaps eventually halted, and most

of the surviving species saved. Some of the most important steps that can be taken are outlined in the Convention on Biological Diversity signed by 156 nations and the European Union at the 1992 Earth Summit in Rio de Janeiro. The convention was the turning point in the awareness of biodiversity as a world issue. Besides speeding up conservation efforts, the convention awakened many tropical countries, where biological diversity is both the richest and the most threatened.

The new approach to biodiversity preservation, uniting conservation and economic development, is not perfect, and it is not yet fully practiced in any country. But it is a promising start. Some of the test projects have succeeded dramatically. They offer a way out of what will otherwise be a biologically barren future. With the world population at six billion and sure to keep on growing rapidly, humanity has entered a dangerous environmental bottleneck. We hope—surely we must believe—that our species will come out the other side in better condition than when we entered. We should make it a goal to take as much of the rest of life with us as is humanly possible.

Spumellarian (Actissa princeps)

Abebe Getahun: Profile

To help his government make decisions about conservation, Ethiopian biologist Abebe Getahun, lecturer at the Department of Biology, Addis Ababa University, is studying carp and other freshwater fish. "Introduction of exotic species, building dams, and channelizing watercourses are all problems that face the Ethiopian freshwater system," he says.

As a young man, Getahun was impressed by the Swedish agricultural agents who came to help local farmers improve seed varieties and the husbandry of farm animals. As a result, he has studied in Ethiopia and abroad so that he can help to improve environments everywhere.

He earned his Ph.D. in biology under a joint program at the City University of New York and the American Museum of Natural History's Center for Biodiversity and Conservation before he returned to Ethiopia.

He plans to write about the fish fauna of Ethiopia and to help form nature clubs to educate local children so they in turn may show their families ways to protect their natural resources.

"Students should be aware of other countries' affairs, not just the problems of their home countries. One country's problems affect other countries. There is no independence, especially when we consider biodiversity loss."

What's This Biodiversity and What's It Done for Us Today?

Norman Myers

Biodiversity primarily amounts to all species on Earth. Our fellow species may be as few as eight million—or as many as thirty million or even one hundred million. We simply do not know how many forms of life there are on our planet. Isn't this absurd? Life is the most startling thing about our planet; as far as we know, this is the only place in the universe that supports life of any kind. Yet we do not know the basic statistics about the richness of life around us. We spend vast sums sending space probes to

Norman Myers has been a consultant on environment and development for more than twenty years.

Mars to see if it has signs of life from billions of years ago. (Conditions on Mars then would have favored life much more than they do now.) If anything turns up, it will be one of the most primitive forms of life such as a slime mold. Yet on Earth we have creatures so sophisticated in form and makeup—consider a giraffe, butterfly, or redwood tree—that we do not remotely realize how complex they really are.

Even more astonishing is that we are telling these other forms of life that Earth is not roomy enough for them and us. Simply by elbowing them out of living space, we are driving them to extinction at a rate between 50 and 150 species per day. If we work with a low estimate that Earth has ten million species in all, this means we are bidding farewell to 0.2–0.6 percent of Earth's species each year. This extinction rate may not seem that high. Nevertheless, it is at least 10,000 times greater than the "background" extinction rate—the natural rate of species' extinction before humans came along with their disruptive activities. Worse, the annual extinction rate is increasing at an alarming rate. By the time today's high schoolers reach old age, Earth may have lost fully half its species—assuming we allow the present spasm of extinction to persist unchecked.

Some people respond, "So what?" After all, even if we lost fifty species yesterday, the Sun still came up on time this morning. What's the big problem? Well, we may be missing many materials that could support our daily welfare. From morning coffee to an evening chocolate snack, we benefit from our fellow species. Here, we exploit crops kept productive by regularly receiving new genes from wild relatives of coffee and chocolate. These wild genes offer resistance to whatever new pests and diseases come along from year to year, helping to ensure good harvests. Each day, we humans use hundreds of products that owe their origin to wild plants and

animals. Conservationists rightly proclaim that by saving the lives of wild species, we may be saving our own. Up to now, scientists have thoroughly investigated only one percent of Earth's 250,000 plant species, and a far smaller percentage of its millions of animal species. Who knows what future benefits the now-unknown portion of biodiversity's genetic library could provide us?

Wild genes support nearly all our crops. Wheat and corn genes in plants from developing countries economically benefit industrialized countries such as the United States by more than $2.7 billion a year. In addition, many foods new to United States consumers stem from wild species. United States supermarkets now feature all kinds of exotic vegetables and fruits. From 1970 to 1985, the number of items available doubled to more than 130, and sometimes increased to as many as 250. By the mid-1980s, this specialty produce, mostly from the rain forests of Asia and Latin America, had become a $200 million-a-year business.

Consider the importance of plant-based medicines. Their commercial value in developed nations topped $40 billion a year in the late 1980s. Two anticancer drugs from the rosy periwinkle plant—native to the rain forests of Madagascar—generate sales totaling more than $250 million per year in the United States alone. According to the National Cancer Institute, tropical forests alone could contain twenty plants with materials for several more anticancer "superstar" drugs. Anticancer drugs derived from plants save around 30,000 lives in the United States each year. The resulting economic benefits amount to $370 billion in lives saved, suffering relieved, and worker productivity maintained.

An economic assessment of tropical forest plants' potential worth worldwide, not only for anticancer purposes, ranges from $420 billion

Moroccan solitary bee *(Panurgus)*

to $900 billion. The total commercial value of plant-derived medicines to developed nations alone could amount to $500 billion during the 1990s. This estimate includes only the commercial value as determined by sales of plant-derived products. The total economic value, which includes the benefits of a reduced incidence of illness, reduced mortality, and so on, will likely be several times as large.

Besides supplying us material goods, Earth's species supply us with many environmental services. These services include generating soils and maintaining their fertility; converting solar energy into plant tissue; helping to sustain the movement of water through Earth's ecosystem; supplying clear air and water; absorbing and detoxifying pollutants; decomposing wastes; enabling such crucial chemical elements as carbon, nitrogen, phosphorus, and sulfur to cycle between the living and non-living parts of Earth's ecosystem; and controlling the proportion of the different gases making up the atmosphere, thus helping

to determine Earth's climate by affecting the rate at which Earth loses heat (through the atmosphere) to outer space. Reducing Earth's biodiversity reduces its efficiency in providing these and other environmental services.

Consider the role of biodiversity in protecting soil. Rainfall runoff erodes the soil of deforested watersheds and carries silt downstream. This silt often accumulates in reservoirs, reducing their capacity for water. This siltation costs the world's economy $6 billion a year in lost hydroelectric power and lost irrigation water. In the past 200 years, the average topsoil depth in the United States has declined from twenty-three centimeters to fifteen centimeters. This has cost the average American consumer around $300 per year in lost water and nutrients. The total cost to the United States of this loss is about $44 billion annually. Worldwide costs of soil erosion are at least $400 billion per year.

Consider, too, the important though little recognized services wetlands such as swamps and marshes provide us: they supply fresh water for our household needs, treat sewage, and cleanse industrial wastes. They provide habitats for commercial and sport fisheries, and recreation sites for canoeing and bird watching. They reduce flood damage by providing locations for flood water to accumulate harmlessly. Louisiana wetlands provide services worth an estimated $6,000 to $16,000 per hectare. At the lower estimate, the current annual loss of these wetlands costs local communities at least $600,000 per square kilometer per year. Similarly, marshlands near Boston have an estimated value of $72,000 per hectare per year, solely for their role in reducing flood damage.

About one-third of the human diet comes from plants that depend on insect pollination to reproduce. Wild bees and domesticated honeybees together pollinate $30 billion worth of ninety different U.S. crops annually, and pollinate many wild plant species.

Finally, note the vital part played by biodiversity in the fast-growing business of "ecotourism." Each year, people taking nature-related trips contribute at least $500 billion to the national incomes of the countries they visit. Much of these ecotourists' enjoyment comes from the biodiversity they encounter. In the late 1970s, the average lion in Kenya's Amboseli Park earned $27,000 per year in tourist revenues, while the average elephant herd earned $610,000 per year. Today these figures are much higher, as many more tourists are visiting the park. In 1994, whale watching in sixty-five countries attracted 5.4 million viewers and generated tourism revenues of $504 million. According to conservative estimates, a pod (group) of six-teen whales at Agate in Japan would earn at least $41 million from whale watchers over the next fifteen years; if whalers killed them, their carcasses would be worth only $4.3 million. In 1970, ecotourism in Costa Rica's Monteverde Cloud Forest Reserve generated revenues of $4.5 million, or $1,250 per hectare; farmland outside the reserve earns only $100 to $300 per hectare. Florida's coral reefs earn an estimated $1.6 billion a year in tourism revenues.

What about an overall reckoning? A team of ecologists and economists has recently attempted a comprehensive estimate of the dollar value of all goods and services stemming from biodiversity. They offer a preliminary and tentative total of $33 trillion per year. In comparison, the world's economy is worth $28 trillion per year.

Consider, too, Biosphere 2, the glass-enclosed artificial ecosystem in the Arizona desert with life-support systems designed to sustain eight Biospherians over a period of two years. (In fact, some air had to be admitted from outside to reduce a potentially dangerous buildup of carbon dioxide in the Biosphere.) The total cost was about $150 million, or $9 million per person per year. Biodiversity has provided these same life-support services to the rest of us for free and, at least so far, without the kind of life-threatening glitches that plagued the Biosphere. If we were charged at the rate of Biosphere 2, the total bill for all Earthospherians today would come to roughly $50 quadrillion ($50,000,000,000,000,000) per year.

Finally, we know so little about the extent and role of biodiversity that we do not know how much we can reduce biodiversity without jeopardizing ourselves. As E. O. Wilson has warned, "If enough species are extinguished, will ecosystems collapse and will the extinction of most other species follow soon afterwards? The only answer anyone can give is: Possibly. By the time we find out, however, it might be too late. One planet, one experiment."

Rotifer (*Trochosphaera solstitalis*)

Spotted ratfish (*Hydrolagus colliei*)

Overconsumption poses a very real threat to biodiversity.
The following seafood guide shows marine biodiversity
at risk largely as a result of our food choices and
other over-exploitation. Reprinted from *Audubon* magazine,
this guide was compiled by Carl Safina, director of
the National Audubon Society's Living Oceans Program.
It draws on research from a variety of sources,
from governmental agencies to environmental groups.

The *Audubon* Guide to Seafood Carl Safina

Species or Group		Background
Sharks (400 species worldwide, including mako, thresher, dogfish, a.k.a. cape shark)		Sharks mature late in life, grow slowly, and produce few offspring. As a result, their populations require decades to recover from intensive fishing. Sharks are often caught for shark-fin soup, sold in China for $90 a bowl. Shark cartilage is now being exploited for "miracle" drugs.
Swordfish, marlins (one species of swordfish; several marlin species, in tropical to temperate seas)		Their impressive size, sleek appearance, and superb hunting skills make these billfishes perhaps the most spectacular sea fishes. But their popularity as pricey steaks is depleting the species.
Shrimps		A wide variety of shrimps come from all over the world, from the tropics to temperate climes. About half are farmed, mostly in the tropics. Shrimp farms pollute and destroy habitat—so much so that the Indian government recently ordered more than 100 farms closed.
Orange roughy		Most orange roughies come from deep waters off New Zealand and Australia. Depleted populations require many years to recover. Orange roughies grow very slowly, taking 20 years to reach spawning age. They can live more than 100 years.
Groupers		Overfishing threatens this large tribe of predominantly tropical species. Fishing in some spawning areas has depleted many populations. Groupers change sex with age, so heavy fishing—which takes most of the old fish—can wipe out an entire sex.
Atlantic groundfishes (Atlantic cod, haddock, pollack, "scrod," yellowtail flounder, monkfish)		These species are mainly caught off Newfoundland, New England, and Europe. After several decades of overfishing and mismanagement, their collapse probably ranks as the world's greatest fishery-management disaster.
Scallops		Two varieties are usually sold: sea scallops, which are from deeper waters, and bay scallops, from shallow East Coast estuaries. Some scallops are farmed.
Salmons (one Atlantic species, six Pacific species)		Nearly all salmon hatch in rivers and grow in the sea, then return to the river to spawn. They are native only to the Northern Hemisphere but have been introduced to South America, New Zealand, and the Great Lakes. Nearly half of all salmon sold is farmed.
Tunas (five major species: bluefin, bigeye, yellowfin, albacore, skipjack)		Almost all large bluefins are shipped to Japan for sushi. There, bluefins are often worth $10,000 to $20,000 each (the record is $80,000), making them one of the world's most valuable animals. Canned "white tuna" is albacore; "chunk light" is yellowfin or skipjack.
Pacific rockfishes (more than 50 species)		Pacific rockfishes are a valuable commodity on the West Coast. Often marketed as Pacific red snapper.

The number of fish—from one (low) to three (relatively high)—denotes the status of the population(s).

An upward arrow means good management; sideways, fair; downward, poor.

The number of hooks—from one (low) to three (high)—indicates the number of other marine animals unintentionally killed in catching this fish, as well as other negative effects of the fishery.

Status	Management	Bycatch & Habitat Concerns
Many populations are declining. Most species on the East Coast are overfished and depleted.	Poor in the Pacific Ocean. Management is fair to good in U.S. Atlantic waters. Almost no management elsewhere.	Moderate to high. Most shark fisheries use longlines or gill nets, which also catch unwanted fishes and creatures such as turtles and marine mammals. Many sharks are killed just for their fins, then dumped.
Overfished and depleted in the Atlantic. Their status is unknown in most of the Pacific.	Ineffective in the Atlantic. Virtually nonexistent in the Pacific. Atlantic marlins may not legally be sold in the U.S., although that has not stopped their decline.	High. Most swordfish and marlins are caught with longlines, which bear thousands of hooks, or in drift nets. Both methods take high numbers of juveniles, sharks, turtles, and some marine mammals.
Plentiful in some regions, depleted in others (such as Mexico's Gulf of California). Their status is not well known elsewhere.	Generally poor in the U.S., and even worse in many other countries. Regulation of farming, the effects of bottom-trawl nets, and bycatch are the main issues.	The highest of any fishery in the world. For every pound of shrimp you buy, an average of seven pounds of other sea life was killed and shoveled overboard. West Coast spot prawns are a rare exception; they are caught by traps that let other creatures escape.
Many populations were severely overfished when the mild-tasting species' popularity soared, in the 1980s.	Poor in New Zealand, where successive populations are being completely fished out. Good in Australia. Nonexistent in international waters.	Significant. The trawls used to catch orange roughies cause serious damage to ocean-floor habitat.
The Nassau grouper is one of the few ocean fishes ever proposed for listing in the U.S. as an endangered species. Many other species are overfished and depleted.	Poor. Many groupers, especially in the tropics, are taken in unregulated fisheries. Management in the southeastern U.S. is improving.	High. Groupers are frequently caught with wire traps, which keep killing even if lost at sea. Bycatch is also high from hook and line. Juvenile groupers tossed back into the sea don't necessarily survive, because pressure changes cause injury when they are hauled up from deep water.
Overfished and depleted, to the point of disrupting fishing communities in New England and Canada.	Poor. Mortality rates are high in many areas, such as the Gulf of Maine. In large areas that have recently been closed to fishing, depleted populations are slowly increasing.	High. The bottom-trawl nets used for these species sometimes entail the highest rates of bycatch of any fishery, except shrimping. The nets also scour the seabed, which degrades the habitat, lowering the potential for recovery in vast areas.
Atlantic sea scallops are overfished and depleted. Bay scallops are having trouble with harmful algal blooms.	Generally poor. Varies on a regional basis.	High. Dredging for scallops takes many other species and severely disrupts habitat on the ocean's bottom. The exception is farmed scallops, which are grown indoors.
Healthy in Alaska; most wild salmon elsewhere are in severe trouble. Several salmon populations are listed as endangered, and many are extinct.	Good in Alaska; poor elsewhere. Salmon are most at risk not from fishers—who are the chief economic force behind their protection—but from logging, agriculture, and dams.	Low bycatch. But salmon farming pollutes, displaces wild fish, and prompts the shooting of predatory seals near farms.
Bluefins are severely overfished. Bigeyes, yellowfins, and albacores are declining in some regions. Skipjack populations are still large.	Poor in the Atlantic, where populations are most depleted. Current management in the Pacific is not adequate to prevent future depletion.	Moderate. Tunas sold in the U.S. must be "dolphin-safe" (no dolphins killed), but many dolphin-netting methods catch juvenile tunas and unwanted species. Bigeyes and Atlantic yellowfins are often taken on high-bycatch longlines. Troll-caught tuna is OK.
Poorly known for many species. Pacific rockfishes are especially vulnerable to overfishing: They can live more than 100 years and may take decades to recover.	Fair to poor. But overfishing of a number of key species is now forcing managers to sharply reduce the catch.	Moderate. Pacific rockfishes are caught with either nets or longlines, both of which also take a number of other sea creatures.

Compiled by Carl Safina, Ph.D., director of the National Audubon Society's Living Oceans Program. In *Audubon* magazine, May–June 1998, Vol. 100, pp. 64–66. Research assistance by Rachel X. Weissman. Copyright © 1998 *Audubon* magazine (To order, call 800-274-4201.)

Species or Group		Background
Snappers		Snappers are a very large, widely distributed group; most live in the tropics or subtropics. The two best known are yellowtail snapper (not to be confused with Pacific yellowtail, called hamachi in sushi bars) and red snapper.
Clams, oysters		Clams and oysters are a big, diverse group living throughout the world; there is a vast array of fisheries and methods for catching them. Many clams and oysters are farm-raised.
Lobsters (two types: American, or "Maine," and various spiny lobster species)		Relatively slow growth and late maturation leave lobsters vulnerable to heavy fishing pressure.
Summer flounder a.k.a. **fluke**		This East Coast flatfish inhabits estuaries in summer and deep waters on the continental shelf in winter.
Halibuts (three species: Atlantic, California, Pacific)		Largest of the flounderlike flatfishes. Atlantic and Pacific halibuts can reach 700 pounds.
Dolphinfish a.k.a. **mahimahi** or **dorado**		Found in offshore waters in the tropics and subtropics, the acrobatic dolphinfish is probably the world's most beautiful fish. Its vibrant colors—yellow, green, and blue—flash in life but fade upon death.
Bluefish		Recreational anglers catch more of this active, voracious species than any other on the East Coast. Bluefish appear subject to large fluctuations in their natural population.
Mackerels (many species, including Atlantic, pacific, Spanish and King)		Mackerels are related to tunas and are found worldwide in warm and temperate seas. Many are important prey for other fish.
Squids		Many species exist worldwide, in habitats ranging from shallow bays to abyssal depths.
Crabs		A very large and diverse group from many regions. They inhabit shallow to deep water.
Striped bass		Abundant, after severe depletion in the 1980s. Half of all fish sold now are farmed, and the share may soon increase to 75 percent. Striped bass farms tend to be relatively benign, because they grow the fish in tanks, where water pollution is controllable and the fish can't escape.

The number of fish—from one (low) to three (relatively high)—denotes the status of the population(s).

An upward arrow means good management; sideways, fair; downward, poor.

The number of hooks—from one (low) to three (high)—indicates the number of other marine animals unintentionally killed in catching this fish, as well as other negative effects of the fishery.

Status	Management	Bycatch & Habitat Concerns
Red snapper is depleted. The status of yellowtail snapper is unknown, but it is probably in fair shape.	Fisheries are unmonitored in most countries. Management in the U.S. is generally poor. Red snapper is over-fished in the U.S., primarily because shrimp nets kill billions of juveniles.	High. Fishing for snappers entails significant bycatch of juveniles and nontargeted species.
Some populations are strong; others are depleted.	The adequacy of management and monitoring varies widely from region to region.	Significant. Taking shellfish often involves serious destruction of habitat on the ocean bottom. But many farmed oysters, especially on the West Coast, are grown on racks, which does not affect the bottom.
American lobsters are overfished, though not depleted. Spiny lobsters are overfished in many parts of the tropics.	Poor for spiny lobsters. Intense fishing pressure takes nearly all American lobsters as soon as they attain legal size, leaving the species vulnerable to one bad year of reproduction.	Low. Most American lobsters are caught in habitat-friendly, low-bycatch traps. In some tropical countries the main bycatch is fishers themselves, who dive with bad equipment and little training; they often get the bends.
Increasing from very low levels.	Good. Strict management is resulting in a recovery, although the summer flounder is still considered overfished.	Moderate. Bottom-trawl nets are hard on habitat and take moderate to high bycatch. There's less bycatch and no habitat damage if the fish is caught on a rod and reel.
Severely depleted in the Atlantic by trawl nets that target fish from the cod family and others. Halibuts are faring better in the Pacific, especially off Alaska.	No regulation in the Atlantic; halibuts are well managed off the Pacific coast.	High in the Atlantic, from bottom-trawl nets, and off California, from gill nets. Little bycatch in Alaska's well-managed longline fishery. Alaska longliners asked for regulations requiring devices to prevent albatrosses from getting hooked—a big plus.
Widespread and abundant. The dolphinfish is fast-growing and highly fecund.	Poor. Virtually no management any-where, even though fishing is intensifying in some areas—largely to make up for the depletion of billfishes and tunas.	Moderate. In U.S. waters, most are caught by anglers using a rod and reel. However, the increasing use of longlines is causing higher bycatch.
Technically overfished. Bluefish have declined from very high populations in the 1980s, because of overfishing and also, perhaps, declines in availability of prey.	Fair. One of the very few fish for which a management plan was enacted before problems began. Regulators are now addressing overfishing.	Low. Most bluefish are caught on a rod and reel.
Most mackerels are in the safe zone. But the king mackerel is overfished in the Gulf of Mexico.	Variable. Some species are well managed, others not. King, Atlantic, and Spanish mackerels have all been overfished in the past. Management has generally improved.	Low to moderate in offshore purse-seine nets and coastal gill nets.
Squids are generally abundant, because they mature fast. High fecundity allows them to withstand heavy fishing pressure.	Highly variable. Squid fisheries are generally well-managed off the East Coast of the U.S. But the lack of management off the West Coast is raising concerns about overfishing.	Low to moderate. Some commercial squid fisheries use nets; others, hook and line. New methods can reduce bycatch because fish behave differently from squid when encountering nets.
Though crabs are generally in good shape, they are suffering from pollution in certain regions. Alaska king crabs are overfished.	Mostly good, particularly for Dungeness crabs. Poor for Alaska king crabs.	Low for most species. But the king and tanner crab fisheries in the Bering Sea entail high bycatch.
The striped bass remains abundant on its native East Coast. On the West Coast, where it was introduced in the 1800s, it is suffering from damage to its habitat.	Good. This is the greatest success story in fishery management, achieved through fishery closures, lower catch limits, and increased protection of juveniles.	Most nets used for wild striped bass entail moderate bycatch. Bycatch is not an issue if striped bass is farmed in tanks.

Deforestation in the Tropics

Robert C. Repetto

Each year, people destroy tens of thousands of square miles of tropical forests. This vast destruction is a huge economic loss to the countries losing their forests and threatens the entire world ecosystem. The direct causes of this destruction are such practices as inefficient commercial logging and the conversion of forests to farms and cattle ranches. However, the fundamental causes of this devastation are unwise policies of both developing and developed nations.

Robert C. Repetto is a visiting Professor in the School of Forestry and Environmental Studies at Yale University.

Currently, active management for sustained productivity occurs in less than one-tenth of one percent of remaining tropical forests. Moreover, in most countries nothing protects forests designated for logging from encroachment by settlers and short-term farmers, who typically burn and clear the forest. Surveys in the Amazon make it clear that deforestation is particularly rapid where roads for logging or other purposes have opened a region.

Burning valuable timber while clearing forests is only the most obvious kind of wastage. The loggers themselves destroy enormous quantities of timber through careless use of equipment and inefficient logging practices. For example, loggers may extract ten percent of the timber in an area; typically this is from mature trees of the most commercially valuable species. The loggers usually destroy at least half the remaining stock, including both immature trees of the valued species and harvestable stocks of varieties that are less desirable commercially. In addition, loggers often keep reentering partially harvested areas to extract more timber before the stands have recovered from the preceding harvest. This inflicts heavy damage on remaining trees and makes it impossible for the forest to regenerate itself.

The economic losses from the destruction of trees that had potential value as lumber are huge. For example, in Brazil the clearing of forest land by burning destroys $2.5 billion of commercial timber every year. In Ghana, where eighty percent of the forests have disappeared, loggers harvested only an estimated fifteen percent of the timber before the land was cleared.

Potential revenues from timber sales are not the only economic losses in deforested countries. Local residents typically use approximately seventy percent of the wood harvested in tropical countries, mainly for fuel. As forests recede, severe fuel wood shortages loom. Other forest resources also become unavailable to local residents. These include animals killed for meat, fruits, oils, nuts, sweeteners, resins, tannins, fibers, construction materials, and a wide range of medicinal compounds.

Apart from the loss of timber and other forest products, deforestation often has a severe environmental impact on soil, water quality, and even local climate. The soil in tropical forests is typically shallow and contains only low levels of nutrients. Heavy equipment easily damages these shallow soils. In addition, heavy tropical rains quickly erode them, or at least dissolve and carry away their nutrients. The increased sedimentation that results from erosion has damaged river fisheries and critical seasonal habitats for fish. Large-scale deforestation interrupts moisture recycling, thus reducing rainfall, raising soil temperatures, and perhaps promoting long-term regional ecological changes. (In intact forests, water from the soil enters the trees' roots and then travels to the trees' leaves, where it then evaporates into the air.)

Moreover, this rapid deforestation poses extreme risks to the world's overall ecosystem. Scientists estimate that the release of carbon dioxide from burned forests to the atmosphere amounts to fifteen to thirty percent of annual global carbon dioxide emissions; thus, this burning contributes substantially to the buildup of greenhouse gases. In addition, the loss of tropical forests is rapidly eliminating the habitat of many plant and animal species. About half the world's species inhabit tropical forests. The ten most biologically rich and severely threatened regions account for 3.5 percent of the remaining tropical forest area; in these ten

32 | 33

Satellite image before deforestation in Rondonia, Brazil, 1975.
This image represents an area of approximately 140 square kilometers.

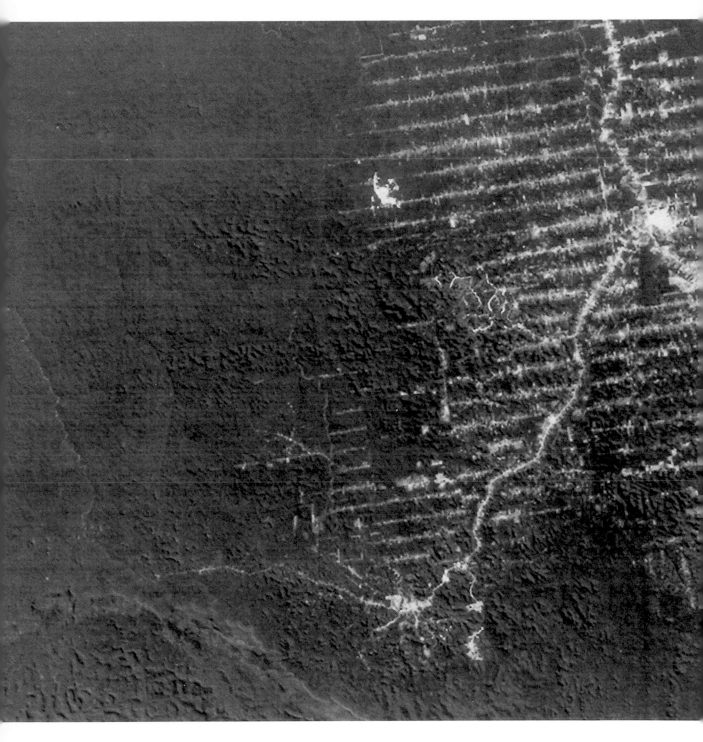

Satellite image showing deforestation in Rondonia, Brazil, 1992.
(Notice the herringbone pattern, which indicates deforestation.) This
image represents an area of approximately 140 square kilometers.

regions, seven percent of all plant species will probably have become extinct by the end of the century, if current trends continue.

Overall, both experience and analysis reinforce the argument that deforestation has not been a path to economic development. In most tropical countries, it has instead been a costly drain on increasingly valuable resources. Moreover, this deforestation need not have happened. It is largely the consequence of poor stewardship, inappropriate policies, and inattention to significant social and economic problems unrelated to the forests themselves.

To begin with, the governments of most developing countries in the tropics have not put an adequate value on their forests. In this region, governments own at least eighty percent of mature forests of the type characterized by a dense upper layer called the canopy. As owners, these governments could receive payments equal to the entire resource value of the forests' timber minus the cost of the labor and equipment committed to managing and harvesting the forests. They could do so simply by charging high enough royalties (fees for cutting rights) and taxes or by selling harvesting rights to the highest bidders. Instead, almost all governments have allowed most of the real economic value of these resources to flow to timber concessions and speculators, who often represent foreign enterprises.

Governments have created huge windfall (unearned) profits for a few individuals— politicians and their relatives and also business people. They have done so by keeping royalties and fees charged to timber-concession holders low, by reducing export taxes on processed timber (for example, plywood) to stimulate domestic industry, and by excusing logging companies from paying income taxes. Moreover, governments have often even failed

to collect the agreed-on fees. Few tropical countries have added the full value of the forest resource to the public treasury—by limiting timber exploiters to a normal rate of profit.

The resulting bonanza atmosphere has sparked timber booms throughout the tropics, drawing both domestic and foreign entrepreneurs— many with little forestry experience—into the search for quick fortune. Under their pressure, governments have awarded timber concessions that cover areas far greater than they can effectively supervise or manage; these areas sometimes extend beyond designated production forests into protected areas and national parks.

While they sacrifice enormous sums in potential forest revenues, governments in the tropics are failing to invest enough in stewardship and management of the forest. For example, in Indonesia nearly half of all trained foresters work in the capital city of Djakarta, which is hundreds of miles by sea from the forests. Those who do get out into the field can often get shelter and transportation only from the people or companies they are supposed to be monitoring—those holding the harvesting rights. Because the foresters are thus dependent on the harvesters, they are less likely to enforce the rules against the harvesters. All this means that forests are not being managed to ensure sustained productivity—even when there are suitably worded concession agreements.

Another harmful type of government policy is the establishment of "perverse incentives" in the very agreements reached with timber companies. The wording of these agreements often discourages any possible interest loggers might have in managing the forests for sustained yields. For example, for a forest to provide a sustainable yield requires intervals of twenty-five to thirty-five years between successive harvests in "selective cutting systems"; "monocyclic systems" (when loggers extract all salable timber at once) require even longer intervals. However, most agreements run for twenty years or less, some for less than five years. Therefore, because concession holders will lose their rights to harvest after this short period, they have no economic reason to care whether the forest maintains its productivity for future harvests.

Sometimes governments charge loggers fees that depend simply on the volume of wood extracted (no matter whether this volume represents high-value or low-value wood). This encourages "high grading," a practice in which loggers take out only logs having the highest value and do so over large areas of forest and at minimum cost. The loggers do not pay for trees they simply destroy but do not take out of the forest; consequently, they have nothing to lose by destroying trees worth less than the per-tree fee being charged. The result of these harmful economic incentives is that loggers often inflict extensive damage on residual stands. Royalties based not on what they extract but on the size of the concession and on the total salable timber it contains would encourage more complete use of the timber within a smaller harvesting area.

Distorted incentives also reduce the efficiency of wood-processing industries. Many countries seek to increase both employment and the value added to forest products domestically by encouraging the export of processed wood products rather than the export of logs. They must provide strong economic incentives to local mills to overcome high rates of protection against the importation of processed wood in Japan and Europe. (Japan and Europe place a high tariff—a type of tax—on imported processed wood. This tariff is added to the price charged by the exporter, which tends to make the total price to the consumer higher

than that for a comparable product produced domestically—that is, in Japan or Europe. To offset the effect of these tariffs, the developing countries exporting wood provide incentives—financial subsidies, perhaps—to the companies making these products. Because these companies have received these incentives, they can set an export price for their products low enough that European or Japanese consumers will still find them competitively priced compared with the domestic equivalents, even after their governments have added a tariff to the total price.) Extreme measures, such as bans on log exports or quotas on exported logs that are based on the volume of logs processed domestically, have created inefficient local industries. Sometimes, in fact, these local processing industries are set up only to preserve valuable log-export rights. (In other words, if these companies produce a certain amount of processed wood products, however inefficiently they do so and however small the market for these products, they can then export a proportional amount of logs, for which they do get a good price.)

Countries sheltering inefficient processing industries can experience heavy economic losses. In the Philippines, each log exported as plywood is worth from $100 to $110 less per cubic meter than it would be if exported without processing (that is, as logs) or as sawed timber. The government sacrifices more than $20 million annually in foregone export taxes to encourage these plywood exports.

Industrialized countries have contributed to—and profited from—these forest-policy problems in the tropics. European and American companies have held interests in logging and processing enterprises, especially in tropical Africa and Latin America. However, Japanese business now heavily outweighs its rivals in the tropical timber trade. The Japanese have shown little interest in the sustained management of their holdings; their highly leveraged operations (operations financed by large loans with high interest rates) have harvested as much as possible as fast as possible to pay off financing charges. Moreover, Japanese firms have participated in the bribery, smuggling, and tax evasion that make tropical timber cheap to import and also deprive exporting countries of much of the value of their resource.

Misguided agricultural policy often abets inadequate forest policy and management. Many countries actively encourage the conversion of tropical forests to other uses. In many states, the laws governing the right of people to live on a particular piece of land allow private parties to obtain title to (legal ownership of) forested land by showing evidence of "improving" it—by clearing away the trees. In the Philippines, Brazil, and elsewhere, recognized rights of occupancy (living on but not owning) or possession (ownership) are awarded based on the area of land cleared. Such provisions often become a mechanism for taking land from the public forest estate and changing it into private property. Those who obtain ownership soon sell their piece of land to larger capitalists, who consolidate the pieces of land to establish private ranches and/or to accumulate speculative holdings (property that they intend to sell for a profit rather than use or live on).

Often, such activities would result in loss of money—if there weren't heavy government subsidies. In the Brazilian Amazon, road-building projects financed by the federal government and multinational development banks have fueled land speculation. Although the Brazilian government has suspended incentives for new cattle ranches in Amazonian forests, supports continue for existing ranches, covering 12 million hectares, that have already cost the treasury more than $2.5 billion in lost revenue.

More general agricultural policies contribute indirectly to deforestation. In Latin America and the Philippines, the combination of the better agricultural land into large, generally underused estates pushes the growing rural population into forested frontiers and upper watersheds. The extreme concentration of landholdings is supported by very low agricultural taxes that make farms and ranches attractive investments for wealthy people; it costs them almost nothing to keep extensive holdings that generate relatively little income.

Subsidized rural credit programs also promote land concentration. Laws specifying the maximum interest rate that banks can charge their borrowers inevitably lead banks to reserve credit for large landholders who have ample collateral (property that the banks can seize if the debtors do not make their payments) and secure titles (strong legal evidence of their ownership of the property that serves as the collateral). The large landholders can thus receive virtually free credit and use it to buy out small farmers, who are unable to borrow the money they need to buy equipment that will increase their agricultural productivity.

In many countries, deforestation has provided a temporary escape valve—a respite from development pressures that can be dealt with effectively only at a more fundamental level. In the Philippines, population growth rates in the forested uplands are even higher than the high national average of 2.5 percent per year; this results in high rates of deforestation and soil erosion. Yet the government has been reluctant to address population control directly or to attack highly unfair patterns of landholding in the lowlands.

A prime cause of the rapid deforestation in the tropics during the 1980s was the exceptionally difficult economic conditions facing most tropical countries. Indonesia's drive to export timber products is a conscious effort to offset its lower petroleum earnings and protect its development program from further cutbacks. Many of the most heavily debt-burdened countries are also those with most of the remaining tropical forests. The 1980s was the first period in forty years when economic growth in those countries failed to outpace the increase in the labor force. Employment in the organized urban sector stagnated and declined; real wages plummeted in the informal urban labor market. Instead of the usual rural-to-urban migration, there was a pileup in agriculture.

Is there hope for improvement? There are, in fact, signs of a new approach to forest policy reflecting increasing awareness of the national and global significance of tropical forests. Many countries are taking steps to capture resource rents at full value. Some governments are now strengthening their forest management capability with the help of development assistance agencies. International interest in tropical forests has bloomed, accompanied for the first time by a willingness to contribute to their maintenance. Several voluntary organizations in developed countries have raised money for debt-for-nature swaps: they buy up a bit of the external debt (debt owed to institutions outside the country) of a tropical country at a discount and then exchange it for a local-currency fund (usually to be managed by a local voluntary agency) that will finance forest conservation programs. Some business groups have also taken an active interest.

Nevertheless, there remains a great deal more that the world outside the tropics might do. Inappropriate consumption of tropical hardwoods (for disposable concrete molds, for example) contributes to deforestation. Some businesses in industrialized countries are still taking part in forest destruction. Development-

assistance agencies are still financing activities that are destructive to tropical forests. The African Development Bank has recently agreed to a project that will run a road through one of the Ivory Coast's few remaining tracts of rain forest and mangrove habitat.

The potential scope of international cooperation to halt the destruction of tropical forests is large. The Tropical Forest Action Plan provides one useful framework. The Montreal Protocol for protection of the ozone layer and a proposed convention to mitigate global climate change could also be powerful mechanisms for international cooperation. Fees levied on chlorofluorocarbons and taxes on fossil fuel and other greenhouse gases in industrialized countries would help to reduce emissions and provide funds needed to carry out national programs formulated under the Tropical Forest Action Plan. Programs that link debt reduction to improved resource management and conservation could be expanded substantially.

New forms of international cooperation would reflect the world's growing awareness that the disappearing tropical forests are not only national treasures but also essential elements of the biosphere on which everyone, everywhere depends.

Biodiversity and Human Health

Francesca T. Grifo

Although the intrinsic value of biodiversity alone justifies its conservation, it is worth noting its many services that are essential for human health. It assures sufficient food and water supplies, keeps populations of disease-causing organisms in check, provides a source of materials for medical therapies, offers models for medical discoveries, and gives warnings of toxins and other environmental hazards. The disruption of ecosystems and loss of biodiversity threaten supplies of the food we eat, the water we drink, the air we breathe, and the medicines we need. Earth's animals, plants, fungi, and microbes are essential to human health and well-being—and to life itself.

Francesca T. Grifo holds an adjunct faculty appointment at Columbia University.

The appearance in 1970 of severely deformed Roseate Tern chicks on Great Gull Island (a research station of the American Museum of Natural History) in Long Island Sound showed for the first time that wild populations could be affected by industrial chemicals. This three-week-old chick was hatched with crossed mandibles. Toxic chemicals were found in the terns and in the fish they ate, providing an early warning of pollutants.

Food

All modern crop varieties were domesticated from wild plants. Food production through modern agriculture still depends on these wild relatives as a source of genetic diversity for the continuing improvement of today's crops. A flow of genes from wild plant relatives is also needed because of the changes in environmental conditions, the need to combat plant pests and diseases, and the growing worldwide demand for greater quantities of more nutritious food. Without sufficient food, chronic (long-lasting or frequently recurring) malnutrition weakens people's immune systems. This can increase their susceptibility to intestinal parasites, hepatitis, malaria, and respiratory infections.

Water

Destruction of wetlands, clearing of watersheds, and poor agricultural management all lead to contaminated water and cycles of flooding and drought. Adequate water quality, quantity, and sanitation are crucial because so many disease-causing organisms are waterborne. When water supplies are limited and/or contaminated with human waste, diseases such as diarrhea, cholera, and dysentery become common.

Disease Ecology

Many agents of infectious disease live in natural environments, where their entire life cycles take place in nonhumans. There are many complex sets of interactions among these disease-causing organisms and the other living components of their ecosystems that serve to keep humans and pathogens apart and hence protect humans from disease. Human-caused damage to existing habitats can change the equilibria between predators and prey and between hosts and their parasites. Ecological changes can modify the behavior of disease organisms. Parasites may switch to humans when their natural hosts become rare or disappear completely. Humans may be exposed to "new" diseases when expansion of farmland or other encroachments bring people into close contact with wild animals.

AIDS (acquired immune deficiency syndrome) is an example of a "new" disease that most likely resulted when a pathogen moved from one host organism to another. Before appearing in humans, the human immunodeficiency virus (HIV), the cause of AIDS, appears to have existed in some African monkeys.

Predators often keep diseases in check by controlling the population of disease carriers. When people or other factors decrease the population of these predators, new diseases emerge and old diseases reemerge. Examples of this are two viral hemorrhagic fevers. These are serious and often fatal diseases whose symptoms are fever, bleeding, a decrease in the number of blood platelets, a seriously insufficient blood supply to major organs, and nerve damage. Two varieties—Ebola and Machupo, each carried by a separate, previously rare species of mice—are among the newest emerging infections that are usually fatal to humans. The Machupo virus appeared suddenly in eastern Bolivia when people converted wild areas to agricultural use; excessive use of the insecticide DDT killed off the wild cats that preyed on the virus-carrying rodents.

An outbreak of hantavirus in the Southwest United States occurred when a combination of changes in land use and in climate eliminated coyotes and other animals that eat rodents. Elimination of these predators led to a tenfold increase in the mouse population. This in turn exposed humans to many deer mice that carried the hantavirus, which causes a severe form of hemorrhagic fever.

In much of the eastern United States, deer no longer have significant predators other than people, whose hunting of deer is restricted near human settlements. The result of this and other ecological changes has been a substantial increase in the deer population. Many of these deer carry the species of tick (*Ixodes dammini*) that may be the sole vector (carrier) of the bacterium (*Borrelia burgdorferi*) that causes Lyme disease—named after Old Lyme, Connecticut, where the disease was first reported. The expanded deer population has carried Lyme disease to suburban areas and into contact with people and their dogs and cats. By 1992, Lyme disease was the most-reported vector–borne disease in the United States. (Lyme disease is a recurring disease affecting many body systems. If the disease is untreated, it can result in arthritis of the large joints, muscle pain, and even nerve and heart damage.) Preserving biodiversity can have tangible benefits in reducing the risk of infectious disease.

Medicines

Humans have turned to nature for pain relief for millennia. Animals, plants, fungi, and microorganisms are a major source of compounds for traditional medicine, over-the-counter remedies, and pharmaceuticals. Traditional medicines are the mainstay of approximately eighty percent of the world's population. Typically, these traditional therapies are derived directly from the forests and other ecosystems that used to surround most human settlements. Today these natural medicine chests are fast disappearing, leaving millions of people without remedies.

Many over-the-counter drugs are also derived from nature. Dioscorides, an ancient Greek, wrote that soaked willow leaves placed on parts of the body relieved aches. In 1827, the active ingredient was isolated and named

aspirin. Today in addition to its use as a pain reliever, it is commonly prescribed for the prevention of colorectal cancer and heart attack. *Ephedra sinica*, an odd-looking gymnosperm, was part of the Chinese pharmacopoeia for centuries as a stimulant and treatment of high blood pressure, asthma, and hay fever. After 1920, pseudephedrine from *Ephedra* became the active ingredient in most decongestants.

Habitat loss and fragmentation, overexploitation of species, introduction of nonnative species, pollution and contamination, and global warming are the mechanisms of species extinction. One quarter of all existing biological species may become extinct within the next thirty or forty years. These animals, plants, fungi, and microbes might have been the sources of new foods and new medicines. However, we will never know what unique properties they possess.

Of the 150 most commonly prescribed drugs in the United States, fifty-seven percent contain at least one major active compound derived from, or patterned after, compounds from nature. Eight of the top ten prescribed drugs—such as the antibiotic amoxicillin, the antihypertensive Vasotec, and the antiulcer medicine Zantac—are derived from natural sources: four from animals, three from fungi, and one from plants. Other drugs prescribed less often include the following:

- Quinine and quinidine come from the bark of the Cinchona tree, native to South America. Quinine is still used to treat forms of malaria resistant to

The armadillo is the only animal to acquire leprosy when injected with *Mycobacterium leprae* experimentally. *Mycobacterium* diseases include several opportunistic infections, such as *M. avium,* that attacks AIDS patients.

synthetic antimalarial drugs. Quinidine is one of the most effective drugs for treating abnormal heartbeat rates.

- Pilocarpine, a drug for treating glaucoma, is isolated from a tropical plant long used in traditional medicine in South America. (Glaucoma is increased fluid pressure within the eyeball; if not relieved, this pressure can destroy the neutrons of the retina, causing blindness.)

- Scientists discovered Capoten, used to treat hypertension, through research on the venom of a species of poisonous snake found in New World rain forests.

- Taxol, from the bark of the Pacific yew tree, found in forests in the American Northwest, may be the most promising medication known for the treatment of ovarian and breast cancer.

Nature provides the raw materials that are essential to the development of new drugs and medicines. We urgently need to find medicines for existing ailments that are so far untreatable or incurable, such as many forms of cancer and heart disease. We need new drugs as new infectious diseases emerge and known diseases such as tuberculosis become resistant to existing drugs. Our physiology is in flux as we enter previously uninhabited parts of the world and encounter pathogens for the first time, become a more global society and carry these pathogens with us across continents, change our diets, live longer, and further modify our environment in other ways bound to have consequences for our health. For all these reasons, we will continue to depend on biodiversity as the source of unique molecules for fighting illness.

Learning from Biodiversity

In addition to the well-known plants and animals that have served as medical models, such as

the guinea pig and the fruit fly, multitudes of creatures (many much, much more obscure) have aided our search for knowledge about biomedicine. With the loss of species, we are also losing diverse organisms that help us understand human physiology and disease.

Bears hibernate but do not lose bone mass. Understanding how bears prevent bone loss during four to five months of immobility (the only vertebrates that can do this) could lead to ways of preventing or treating osteoporosis. Osteoporosis (reduction in bone mass that increases the porosity and fragility of bones) is now a largely untreatable condition that afflicts 25 million Americans; it results in 1.5 million bone fractures and 50,000 deaths annually. In addition, bears do not urinate for months during hibernation. Humans, by contrast, would die after a few days if unable to excrete their urinary wastes. Research on bears might lead to effective treatments for kidney disease. People with chronic kidney failure are now totally dependent on a machine to take over kidney function and purify their blood. Renal failure in the United States alone costs $7 billion annually.

Sea squirts are ocean-living animals that are primitive relatives of vertebrates; they have spinal cords but no internal bony skeleton. One type, the sea grape, is the only animal besides humans to form stones naturally in its kidney-like organs. By studying the formation of both uric acid and calcium oxalate stones in these creatures, scientists are trying to understand the mechanisms by which kidney stones and gout develop in humans.

The horseshoe crab—one of the most ancient animals still living—has one of the largest and most accessible optic nerves and the largest photoreceptors of any animal. Scientists have learned much of what we know about how the eye and the brain communicate from the study of this creature.

The armadillo—a mammal notable for its armor plating—is the only animal to get leprosy when injected experimentally with the *Mycobacterium leprae. Mycobacterium* diseases include several opportunistic infections that attack AIDS patients.

Miner's Canaries

In previous centuries, workers in deep mines took canaries down with them into the shafts. When exposed to poisonous gases which often exist in mines, the canaries quickly died. Seeing this, the miners could leave the mine in time to save themselves. Today, some organisms may be so uniquely sensitive to specific assaults on the environment that they serve as our "miner's canary" or indicator species. Other, less-sensitive organisms can act as indicators of environmental toxins if they are at the top of the food chain. Each organism accumulates the toxins in the bodies of all the animals it eats. Feeders at the top of the food chain thus accumulate comparatively large quantities of environmental toxins. This can cause them to suffer harmful effects that we can notice and then tie to the particular toxin.

In 1970, severely deformed chicks of the endangered roseate tern appeared on Great Gull Island, a research station of the American Museum of Natural History. Their crossed mandibles and deformed legs, feet, eyes, wings, and feathers showed, for the first time, that industrial chemicals can affect wild populations. Toxic chemicals were found in the terns and in the fish they ate, providing an early warning of dangerous pollutants in Long Island Sound.

Conclusions

Earth has experienced five mass extinction episodes over the last 500 million years. These resulted from terrestrial and extraterrestrial events. We are now in the midst of the sixth mass extinction, as species vanish in a time span of mere decades rather than millions of years.

We humans are the sole cause of this sixth mass extinction. Our population is now roughly 6 billion—and increases by approximately eighty to ninety million each year. Overconsumption of our natural resources, resulting from this rapid population growth, creates environmental problems. The per capita consumption rate of the one-fifth of the world's people living in the developed nations is ten to one hundred times that of the four-fifths of the world's people who live in developing countries.

For the first time in the history of the planet, we are altering the basic chemistry, physics, and physiology of Earth. We are changing the atmosphere, the oceans, and the land. We are overexploiting natural resources, destroying and degrading the habitats of animals and plants, introducing alien species into new environments, polluting and contaminating the environment with toxic substances, and changing the climate of the globe.

Biological diversity is the key to the preservation of life on Earth. It encompasses the immense range and variety of life-forms— the differences within and between gene pools, species, populations, and entire ecosystems. It sustains and supports all living things, from microbes to humans.

That human health and well-being depend on the conservation of biodiversity is still largely unappreciated. Biodiversity conservation requires bold new thinking about the way we manage Earth. Every policy and action must recognize that Earth's resources are finite and that nature's components are unique and, once lost, irreplaceable.

Kevin Browngoehl: Profile

A pediatrician by training and an outdoorsman by passion, Dr. Kevin Browngoehl has become a leading advocate for the protection of biodiversity. You may find him standing at the side of an elected official during a press conference, writing letters to his fellow physicians, speaking to a group of medical students about the rain forest, holding one of his young patients, or encouraging his or her parents to send empty medicine bottles to Congress to emphasize that many medicines come from nature. Says Browngoehl, "It is a natural extension of someone who is interested in health care and preventive health care to look at the big picture."

Dr. Browngoehl's strong advocacy for preserving Earth's biodiversity stems in part from his experience with children suffering from leukemia. Not long ago, this illness was inevitably fatal. Today, more than seventy percent of children with leukemia are cured—in part due to treatment with a potent drug discovered in the rosy periwinkle plant, which is native to Madagascar. Concerned that weakening the Endangered Species Act (ESA) might result in the loss of many potential drugs, Browngoehl says, "If we continue down the pathway of a short-sighted policy to the ESA, then future generations will not be as fortunate."

Biodiversity: What It Is and Why We Need It

Paul R. Ehrlich and Simon A. Levin

Biodiversity surrounds us. It even lives on us and in us. It touches us in many ways. Yet few people fully appreciate how much we depend on other life-forms. These life-forms are working parts of the ecological systems (ecosystems) that give us indispensable goods and services.

Consider, for example, our food supply. Worldwide, we eat hundreds of species of plants, animals, fungi, and other organisms, either directly or as ingredients in other foods.

Paul R. Ehrlich is Bing Professor of Population Studies in the Department of Biological Sciences at Stanford University. Simon A. Levin is George M. Moffett Professor of Biology at Princeton University and was the founding Director of the Princeton Environmental Institute.

The species we feed on in turn depend on many other species. Grasses are food for cattle. Bees pollinate apple trees. Ladybird beetles devour scale insects that are attacking orange orchards. The trees, shrubs, herbs, and grasses on hillsides receiving rain hold the soil in place and permit a steady flow of irrigation water to farm fields downstream. The tuna in our casserole or sandwich feed on smaller fish, and the smallest fish in this food chain feed on microscopic planktonic organisms. Trees growing in the world's forests absorb carbon dioxide from the atmosphere, slowing global warming and helping avoid a rapid climatic change that could prove catastrophic for agriculture. At the core of the global ecosystem are many species of micro-organisms that provide essential nutrients for more complex life-forms. Some accomplish this by decomposing animal waste, others by "fixing" atmospheric nitrogen gas.

Food production is only one of many benefits we gain from biodiversity. Some bacterial species decompose toxic wastes. Trees and other plants give us shade, and they cool and humidify the air through the evaporation of water from their leaves. Harmless species of bacteria colonize our intestinal tracts, providing us some essential nutrients and helping to protect us from an invasion by harmful bacteria. A variety of predators help protect us from disease by eating insects and other animals that are carriers of disease-causing microbes. Biodiversity also provides us a variety of ecosystem goods, besides all our foods. For example, many of our medicines either come directly from organisms or are synthetic variants of substances first isolated from organisms. Biodiversity also contributes a host of industrial products ranging from timber to oils. Finally, it is a major source of the beauty and enchantment we find in the world.

What is Biodiversity?

The term biodiversity has a variety of meanings. Most often biodiversity means species diversity. Estimates place the number of species of plants, animals, fungi, and microbes at somewhere between 10 million and 100 million different species. As most educated people now know, species diversity is especially high in the fast-disappearing tropical rain forests. Sometimes, biodiversity refers to the diversity within functional groups (also called guilds) of species. A functional group consists of the different species that play the same role in a particular ecosystem. Examples of such roles include predators at the top of food chains, nitrogen fixers, and decomposers. Biodiversity also can refer to the genetic differences (and to the corresponding bodily and/or behavioral differences) within a population of a particular species. Finally, biodiversity sometimes refers to the number of geographically separate populations of a particular species.

By focusing only on the number of species, we fail to see the full importance of biodiversity. In the face of a changing environment, biodiversity (genetic diversity) within a particular species helps it survive or evolve into one or more new species: the greater a population's genetic diversity, the greater the odds that some of its members will thrive under the new conditions. A species' survival can also depend on the distribution of its members over Earth's surface. If an entire species consists of a single population, extinction of that population— perhaps caused by a localized environmental change—would be extinction of the entire species. In contrast, a species could survive the extinction of one of its populations, if it had populations that survived at other locations. Current concern about the increasing rate of species extinction may obscure a perhaps more severe problem: many species now consist of dangerously few populations.

Above the species level, we observe bio-diversity in the variety of biological communities and ecosystems. (Biological communities are groups of species living together. Ecosystems are biological communities plus the nonliving environments with which they interact.) The structure and functioning of an ecosystem depend on the diversity within each of its functional groups (guilds). The more species making up each functional group, the better the group carries out its ecological role. It is also more likely to survive and continue to fulfill its role if one of its species becomes extinct.

Biodiversity's Services to Ecosystems

We can describe an ecosystem by listing its living species and describing its nonliving environment, or by describing the efficiency with which it carries out various chemical processes. Such processes include causing critical chemical elements (for example, carbon or nitrogen) to move between the living and nonliving parts of the ecosystem. They also include producing living matter (biomass), or rendering toxic materials harmless. The challenge facing us is to learn the role and importance of each species in a particular ecosystem.

We know that in some ecosystems, a particular species plays a keystone role in keeping the ecosystem functioning. Loss of such a key species can drastically affect the ecosystem. For example, the near extinction of the California sea otter allowed the population of one species it preyed on—the sea urchin—to increase dramatically. The sea urchins in turn overgrazed on the kelp beds just offshore. As the kelp beds declined, the shoreline suffered increased wave erosion, and fish living near the shore lost much of their food supply. Fortunately, a program of protecting the sea otter caused its population to increase. This reduced the sea urchin population. The kelp

forests grew back, and the ecosystem largely regained its original state.

Numerous examples of keystone species provide only a small insight into ways the different species of an ecosystem interact. Even more important than keystone species are species belonging to certain essential functional groups of species, for example nitrogen-fixing bacteria. Without representation of such an essential functional group, the entire ecosystem would collapse.

Strong parallels occur between individual species and functional groups of species, including the risks associated with loss of diversity. Each species consists of many different individuals. All these individuals are direct descendants of the same series of evolutionary ancestors. Consequently, they have many bodily and behavioral adaptations in common. Nevertheless, each individual organism has its own unique features. When diversity diminishes within a species, the species can survive for a time. However, this decrease in diversity jeopardizes its long-term survival. It has lost some of its genetic capacity to adapt (through biological evolution) to any change in its environment. Its population is less likely to contain individuals with genes that cause them to be adapted to the new environment.

In a similar vein, each functional group ordinarily consists of many species that play a similar role in their ecosystem (often because they are close evolutionary relatives). Nevertheless, each species in a functional group has its own unique biological adaptations. When a functional group loses only some of its species, it can continue to carry out its ecological functions for a time. However, it will not do it as efficiently as before, thus making life more difficult for the other species in the ecosystem. Moreover, it becomes more likely that the entire functional group will become extinct if the environment changes. Because the group contains fewer species, it is less likely to contain any species suitably adapted to the changed environment.

Biodiversity plays an important role in maintaining our life-support systems, beyond providing us food, fiber, fuel, and pharma-ceuticals. Photosynthesis, cellular respiration, and the burial of organic materials serve to maintain an oxygen and carbon balance that permits us to live: our lives, and the lives of most other organisms, depend on the production of energy through cellular res-piration. This life process requires the intake of atmospheric oxygen and produces carbon dioxide as a by-product. All this required oxygen is the product of photosynthesis by plants and one-celled photosynthetic organisms. This same photosynthesis removes carbon dioxide from the atmosphere, helping to keep Earth's climate from becoming too hot. (Carbon dioxide is a "greenhouse" gas that absorbs infrared radiation—radiant heat—that would otherwise pass from Earth's surface into the surrounding space.) In the upper atmosphere, ultraviolet radiation from the sun causes the ordinary form of oxygen to change to ozone. The ozone in turn absorbs incoming ultraviolet radiation, shielding humans and other life on Earth's surface from its potentially deadly effects. Finally, the burial of organic materials by many species of organisms enriches the soil. This aids the growth of plants, which in turn aids the removal of carbon dioxide from the atmosphere through photosynthesis. From all this it readily becomes apparent that any changes in vegetation (or in the population of "burying" organisms) could lead to drastic ecological changes that could jeopardize human life. These changes could include a reduction in the ultra-violet protection provided by the stratospheric ozone layer and severe global warming.

How Important Are the Various Components of Biodiversity?

Studies that merely count species do not reflect that some species are more important than others. Climate change, land degradation, acid rain and snow, and a general accumulation of toxic substances could greatly change Earth's environment. How can we predict which species or groups of species will be most important in such a future world? Ecologists are just beginning to crack this problem. Basic theoretical and field research is illuminating the importance of biodiversity to a variety of ecosystems and is showing how to quantify this importance. Yet we must accelerate and extend this research. Meanwhile, we must preserve not only particular populations but also the critical habitats that are the sites of the greatest biodiversity. Finally, and perhaps most important, we must recognize our major role in the current assault on biodiversity. The chief threats to biodiversity are our rapidly increasing population, our increasing per capita consumption of energy and other resources, and our needless use of environmentally harmful technologies. We can preserve biodiversity only by reversing these trends.

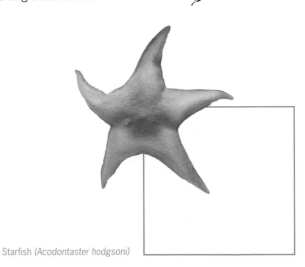

Starfish (*Acodontaster hodgsoni*)

Hot Spots

David Ehrenfeld

Eight friends were seated around the dining room table, two bulging photo albums at the ready, about to start a round of that curious adult version of show-and-tell known as "Our Summer Vacation."

Donna and Stanley had begun their album with pictures of Stanley delivering a lecture to clinical psychologists at the Catholic University of Campinas, in the state of São Paulo, Brazil. This was followed by page after page of lush, colorful South American scenery: snapshots of the mighty Iguazú Falls taken

David Ehrenfeld is Professor of Biology at Cook College, Rutgers University.

from the river below; towering walls of green vegetation along Rio Paraná; a thick growth of native trees overtaking the imported plantings left behind by the celebrated botanist Moises Bertoni at the remains of his agricultural experiment station in the Paraguayan province of Alto Paraná. Then there were pictures of Stanley and Donna with their Brazilian hosts, flanked by stately palms; pictures at their friends' beautiful beach house in Ubatuba; pictures of costumed samba dancers; and more pictures of scarlet and blue macaws, gaudy toucans, purple bougainvillea flowers, brilliant butterflies, and the rich magnificence of one of the last remaining patches of Brazil's Atlantic coastal forest.

The album provided a breathtaking display of natural and human exuberance painted in a profusion of vivid reds, blues, yellows, and, above all, greens. I could almost smell the many fragrances borne on the tropical breeze and hear one of my favorite sounds, the chattering of parrots far above in the treetops.

As the pages turned, I thought of the phrase hot spots, that graphic term for those patches of exceptionally high biodiversity—many of them in the tropics—that receive the lion's share of attention from naturalists dedicated to the preservation of endangered species. I thought of the conservation argument that says if we want to preserve life, look first at the places with the greatest biodiversity—where an acre can yield 100 species of trees instead of a handful, and more species of insects than most of us can imagine. This understandable preoccupation with hot spots reminds me of Sutton's Law, named after the accomplished bank robber Willie Sutton. When Sutton was once asked why he robbed banks, he gave the now classic reply, "Because that's where the money is." The biological "money" is in the hot spots, it seems.

My thoughts drifted back to a much-discussed article in Science that analyzed the geographic distribution of endangered species in the United States. The authors claimed that many endangered species of both plants and animals are grouped together in a few hot spots of diversity—Hawaii, southern California and Arizona, southern Appalachia, and parts of the southeastern coastal states. From this, they concluded that our endangered species can be protected most efficiently by spending conservation efforts and funds in these few key areas. I had been bothered by this article when it first appeared, and was still bothered by it as I looked at Stanley and Donna's pictures. But it was Joan's and my turn to show vacation photos.

Our album was open to a picture of a small, twin-engined Otter airplane disembarking thirteen warmly dressed people and a pile of waterproof bags, canvas Duluth packs, and partly disassembled canoes onto a flat expanse of gravel. Everything in the foreground lies in the gloomy shadow of early twilight—but a low, treeless hill in the background, yellow-brown in color, is brilliantly illuminated by the summer sun. Nothing in the picture is green. Not a single plant is in evidence. "It's very bleak," Judy remarked. "It looks like a desert."

It had felt bleak, I remembered, especially when the little plane, its tundra tires churning the lichen-covered stones, took off, climbed steeply out of the narrow Soper River Valley, and disappeared over the rounded hills of the Meta Incognita Plateau as it headed back to Iqaluit, once known as Frobisher Bay. We had been left at the upper reaches of the only canoeable river on Canada's vast Baffin Island, north of Hudson Strait, in the eastern part of the newly designated Inuit territory of Nunavut. At eight in the evening, the Arctic summer sun was still fairly high in the sky but hidden behind a hill.

A chill wind was blowing. Several members of our group glimpsed a wolf on the other side of the river. We found a partially sheltered spot off the gravel bar and pitched tents in the stunted vegetation while our Maine guides, Randy and Dwayne, cooked dinner on two Coleman stoves. That night we were awakened briefly by the Sun shining on our tent at 3 a.m.

The next page of the album revives my spirits; it shows us two-thirds of the way up Mt. Joy, with the Soper River winding through the valley below. "Everything is so brown," somebody comments. "Wait a bit," I respond, knowing what's coming in a few pages. Meanwhile, here is a shot of Mark showing off a hefty Arctic char, kin to trout and salmon and the one species of fish in the freshwater part of the Soper or any other river on Baffin Island—a landmass that stretches farther than the distance from New York to Chicago. Char are carnivorous: the young eat tiny shrimp and mosquito larvae that are among the few invertebrates in the river, while the older char eat the younger ones. Not exactly an ichthyological hot spot, I muse, as the page turns.

Next, Joan and I are shown running the first big set of rapids, which Randy has dubbed "Gatorade." The page turns again, and suddenly everyone crowds around to get a better view. There are the thundering, 95-foot falls of the Cascade River, a tributary of the Soper, with a giant block of ice miraculously clinging to the adjacent cliff. In the foreground is a patch of "mountain cranberry," dotted with last summer's tasty red berries. The only trees that grow on Baffin, shrubby willows and birches, most less than ten to fifteen inches high, cover the low slopes of the Cascade valley. "How beautiful," Donna says, and I nod; when we were there I had told Joan that this was the most beautiful place I had ever seen. The

Canadian geographer J. Dewey Soper, the first outsider to set foot in this valley with its tumultuous river and falls, called it "a fascinating commingling of savagery and softness." Another photo shows a dry, upper hillside above the valley. Close-up photos reveal the mosses growing between the stones, with every rock sporting a multi-colored array of encrusting lichens.

The page turns and there we are hiking back from the falls to our third campsite. We see the tundra ponds surrounded by low, green vegetation, looking, if you squint a bit, like an aerial photo of a New Jersey salt marsh in June. I thought I saw tiny fish in one of those ponds, but when I brought Dwayne, a fisheries biologist, to look, he pronounced them shrimp. Then I got down close to them and noticed that they were swimming on their backs, just like the fairy shrimp that my teacher, Archie Carr, once took me to see in a temporary pond in north Florida.

Farther down the Soper River, at the fourth campsite, Joan and I stalked a caribou. This is easier than it sounds, in spite of the absence of cover—perhaps they don't see well, or perhaps they are unafraid and curious. The picture shows the caribou, looking too small for its ridiculously large antlers, feeding on plants in a gully about thirty feet from us. We could hear the crackling noises it made as it chewed. When we got back to camp, Randy, who tracks and studies bears for the state of Maine, pointed out a sandy patch where the ever-present caribou tracks were overlaid with the surprisingly big prints of a solitary wolf.

A few miles away, Randy directed us inland to the only two stands of real trees, willows, on Baffin Island. They grow in adjacent patches of several acres each. Some of the willows are as tall as twelve feet. Dewey Soper found and photographed a similar patch—now dead—not far away in the valley. In this part of the Canadian Arctic, the treeline is 400 miles to the south, in northern Quebec. Soper was very excited to come across actual trees that stood up from the ground rather than forming low, spreading mats. Joan thought that the location of the patches in hollows next to sheltering hillsides allows deep snow to accumulate, protecting the willows in winter from the freezing and drying Arctic winds. A photo shows Joan in the midst of the amazing forest. Only a few of the skinny trees are above her head. Another picture shows me beside one of the two 6-foot willows that Joan and I found next to a small stream in a steep-sided little valley several miles away. Possibly no one else has ever seen these two trees. You can see that I am as proud as if I had discovered a giant sequoia anchored in the permafrost.

More exclamations from the people around the table. We have come to the pictures of Arctic flowers blooming in what Soper described as "unrivaled loveliness… in the briefest, though the fairest summer season in the world." Joan has marked each plant with its common and scientific name. Here is a lowland wet spot covered with 8-inch stalks of the cotton grass sedge *Eriophorum angustifolium*, its cottony tufts of plumed seeds waving in the breeze (Figure 1). Not in the photo are the billions of seeds drifting by in the sky above the ridge tops, sparkling in the light as if the air were filled with tiny diamonds. On the top of the next page are the ubiquitous, showy purple flowers of the Arctic willow herb *Epilobium latifolium*, and underneath, a rounded cluster of yellow mountain saxifrage, *Saxifraga aizoides*, looking like an arrangement in a florist's shop. Then comes a picture of mountain avens, *Dryas*

52 | 53

Figure 1: Cotton grass sedge, *Eriophorum angustifolium*

integrifolia. Here is a clump of white heather, *Cassiope tetragona*, nearly finished flowering in late July—a few white blossoms can still be seen among the seed capsules. The major study of Baffin vegetation identifies only 189 species of flowering plants on the entire island, an area larger than the British Isles. In our seven days on the river, Joan found and identified fifty-four species, a significant part of the flora.

The album ends with pictures of some of the beautiful and vivacious Inuit children who accompanied us everywhere while we were in Kimmirut, formerly Lake Harbour, on the southern coast (Figure 2); a picture of three skulls of polar bears—killed in self-defense—resting on top of a shed; and a last view of the Meta Incognita peninsula, brown, lifeless-looking, and laced with streams and narrow patches of snow, as we flew over it on our way back to Iqaluit. Joan closed the book. I could see that the people around the table were moved by what they had seen. However, there was no way that Joan and I could convey the depth of our feelings about this place: so vast and unforgiving, yet so intimate and welcoming—so poor in species diversity, yet so vibrant with life. What we encountered on Baffin Island, and what made it magical, was that increasingly rare quality of nature not yet vanquished, an amazing and nearly pristine ecosystem still in place, still working. It is not a hot spot; it is a different sort of evolutionary pinnacle: nature fully equal to the terrible rigors of a harsh climate.

Why, I wondered, has species diversity become a god to conservationists? The wealth of species that adorns the tropics is one of the greatest glories of this planet. It must be protected. But wealth of species is only one way of reading nature; there are many others. A preoccupation with hot spots can absorb too much of our energy and concern, leaving little for the rest of the natural world. Unspoiled places such as the valley of Soper, poor in species as they are, have their own kind of grandeur and glory, a living presence that surely makes an equally powerful claim on our care and our affection.

Figure 2: Inuit children in Kimmirut (formerly Lake Harbour), Baffin Island, Canada.

Pleasant Village Community Garden, East Harlem.

The Green Guerillas,
New York City John Thomas

In 1974, Liz Christy, a Greenwich Village artist, cleaned up a trash-filled vacant lot on the Lower East Side and turned it into a productive community garden, now known as the Liz Christy Garden. Soon after, she and Bedford-Stuyvesant resident Hattie Carthan formed the Green Guerillas. One of their early restoration activities was to toss seed-filled water balloons over fences into the city's abandoned lots.

The Guerillas now have 800 members and have helped neighborhood groups create and maintain 1,000 urban gardens over the last twenty-five years. In the past three years, as the gardens have become threatened with destruction by the City, the Green Guerillas has switched its focus from horticulture technical assistance and has been organizing coalitions of gardeners to lobby for their gardens' future.

The Liz Christy Garden at the border between the East Village and the Lower East Side grows more than 1,000 different species. It has cactus and moss collections and operates a public learning center. In the East Village, the Green Oasis Community Garden has a mini-arboretum containing fifty tree species and varieties, and offers gardening lessons for

Demolition in progress.

children. It has also built wetland and pond habitats as part of its urban garden environment. In the South Bronx, gardeners at the Rancho Boricua garden grow traditional medicinal and culinary herbs from Puerto Rico. In Brooklyn's Bedford-Stuyvesant, the Hattie Carthan Memorial Garden cultivates an entire city block and, with the Magnolia Tree Earth Center, provides expert assistance to urban growers.

In addition to "greening" neighborhoods and providing tranquil oases of biodiversity within the city, the Green Guerillas also collaborate with the group Just Food on the City Farms Program, which helps gardeners distribute produce to local soup kitchens. By using compost-rich, low-artificial-chemical growing methods, New York City gardens do not generate the environmental pollutants

associated with conventional fertilizer- and pesticide-intensive agriculture. This locally grown food also comes without the packaging and transport costs of distant-grown food. According to the Urban Agriculture Network, city gardens now account for fifteen percent of world food production.

As urban gardening continues to plant roots in vacant lots across the country, American city dwellers are coming to appreciate both the locally grown produce and the pleasures of tending the soil in their own "backyard."

56 57

Roughtailed stingray *(Dasyatis centroura)*

What Have We Lost, What Are We Losing?

Peter H. Raven

Whether we realize it or not, we are entirely dependent on the plants, animals, fungi, and microorganisms that share the world with us. They feed us. They provide many drugs and other products that determine the quality of our lives. They promise us sustainable productivity—productivity that Earth can support indefinitely—of food, materials, and sources of chemical energy. Unfortunately, however, an alarming proportion of these species faces extinction, with possibly dire consequences for us.

Peter H. Raven is Director of the Missouri Botanical Garden and Engelmann Professor of Botany at Washington University.

Earth formed—along with the Sun and the rest of the solar system—about 4.5 billion years ago. Life first appeared on Earth about 3.5 billion years ago. Since then, countless millions of species of living things arose through evolution, had their "moment" in Earth's history, and then became extinct. Normally this extinction resulted from their gradual replacement by newly evolved species with better biological adaptations. However, five mass extinctions (also called catastrophic extinction events) have occurred, during which many of Earth's living species became extinct in a short period.

The most famous of these mass extinctions occurred at the end of the Cretaceous Period, 65 million years ago. Earth then lost about two-thirds of its species, including, most notably, the dinosaurs. (Birds belong to one group of dinosaurs, so, in a sense, the dinosaurs are still with us.) Geological evidence suggests that this mass extinction may have resulted from a catastrophic event that covered Earth in a cloud of dust for a substantial period. While this cloud persisted, Earth's surface received little sunlight. (Geological evidence reveals that an asteroid then collided with Earth, striking its surface near today's Yucatán Peninsula in Mexico. Fine dust particles of a chemical composition typical of asteroids occur in a thin band found worldwide in rocks of the same date, 65 million years ago.) Screening Earth's surface from sunlight would have devastated Earth's plants. This in turn would have devastated the living systems that maintain our planet, driving an estimated two-thirds of all species that existed at that time to extinction. As with the earlier mass extinctions, it took several million years of evolution to restore Earth's ecological health and rebuild its biodiversity. The number of species on Earth has gradually increased over the past 65 million years, from perhaps 500,000 species following the presumed asteroid impact to an estimated 7 to 10 million today.

Today, Earth may be undergoing a new mass extinction. Sadly, up to a quarter of the species on Earth may become extinct or be on the way to extinction by the end of the first quarter of the next century, and up to three quarters of the species may be extinct or endangered by the year 2200. Unlike all earlier mass extinctions, however, the current one appears to result from the actions of a single biological species: *Homo sapiens,* the human species.

In spreading throughout the world, our species, which first appeared about several hundred thousand years ago, has been responsible for the extinction—by hunting—of such familiar prehistoric mammals as mammoths, cave bears, and saber-toothed cats. However, our harmful effects on Earth's ecological health greatly increased with the development of agriculture in several widely scattered centers, starting about 10,000 years ago. Agriculture enabled the human population to increase greatly—at the expense of many other species. The development and expansion of human civilization resulted in much that is praiseworthy. Nevertheless, our sheer numbers, our levels of consumption, and many of our other current practices are destroying much of the global ecosystem in a way that is clearly unsustainable.

Finally, Earth's human population is increasing by an estimated 80 million people annually. Not only does this rapid population growth accelerate the depletion of crucial natural resources but it also increases the risk of fa-mine and social and political instability around the world. Each year we are cutting and burning 1.5 to 2 percent of Earth's existing rain forests. Erosion, road building, urbanization, pollution,

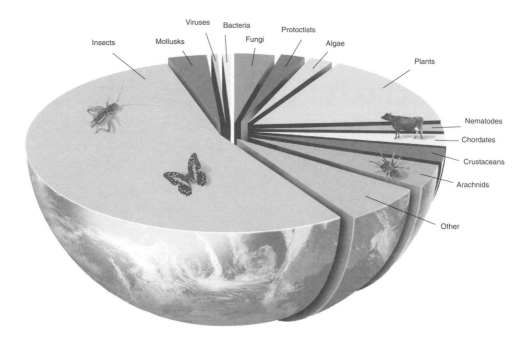

Insects Mollusks Viruses Bacteria Fungi Protoctists Algae Plants Nematodes Chordates Crustaceans Arachnids Other

and other harmful environmental consequences of human activity are costing us an estimated 25 billion tons of topsoil annually; this is equal to all the topsoil on Australian wheat lands.

Every point on Earth's surface, from the frozen expanses of Antarctica to the open ocean, constantly receives artificial chemicals produced by human technology. Many of these chemicals have had drastic effects on Earth's ecosystem. For example, various chlorofluorocarbon (CFC) gases once widely used in refrigerators and air conditioners (but now almost completely banned by international agreement) leaked into Earth's lower atmosphere. Making their way to the stratosphere, where they are long lasting, they then cause a chemical reaction that breaks down Earth's protective ozone layer to oxygen. This ozone layer shields us from harmful ultraviolet (UV) radiation arriving from the sun. (Earth's early atmosphere contained no oxygen and therefore had no protective ozone layer. For that reason, the first forms of life could survive only in ocean water deep enough to screen out UV; UV radiation breaks down nucleic acids

and proteins and causes disruption to life. Eventually, organisms that carried out photosynthesis arose through evolution. The oxygen produced as a by-product of photosynthesis accumulated in Earth's atmosphere, where some of it changed into a protective layer of ozone. Life was then able to move to shallow ocean water and then—about 430 million years ago—onto land.) The recent decrease in the concentration of ozone in the ozone layer has increased the intensity of UV radiation reaching Earth's surface. Among the harmful consequences of this has been an increased incidence of skin cancer.

Since the end of World War II in 1945, the proportion of carbon dioxide in the atmosphere has increased by roughly twenty-five percent; it has increased from 300 to 370 parts per million. This increase results from the rising rate at which we are burning wood and such fossil fuels as coal, natural gas, gasoline, aviation fuel, heating oil, and diesel fuel. When these materials burn, their carbon combines with oxygen from the atmosphere to produce carbon

Figure 1: Proportion of the approximately 1.5 million described species in the major groups of organisms.

dioxide, which escapes into the atmosphere. The increased buildup of atmospheric carbon dioxide is a matter of great concern because carbon dioxide is a so-called greenhouse gas. That is, carbon dioxide absorbs radiant heat (infrared radiation) that would otherwise be lost to outer space. This absorbed radiation changes into heat that increases the atmosphere's temperature. The scientific community has now agreed that Earth's average temperature is increasing—the so-called global warming—because of this increased trapping of infrared radiation emitted from Earth's surface. Significant global warming could greatly alter the world's climate patterns, with possibly drastic consequences for agriculture and for many of Earth's living species. In addition, a higher global average temperature could cause increased melting of ice that has accumulated in Earth's polar regions. This in turn would cause the world's sea level to rise. As a result, many islands and low-lying regions of continents would be under water, again with disastrous consequences for the species—including humans—living in those ecosystems. Other gases, including methane, nitrogen oxides, and CFCs, also enhance the greenhouse effect and are wholly or partly produced as a result of human activities.

How have these human-caused environmental changes affected Earth's biodiversity? We can all point to well-known extinctions caused by human activity. For example, overhunting caused the Carolina parakeet and the passenger pigeon (Figure 2) to become extinct in the first part of our century. However, we are only beginning to develop techniques for estimating the current rate of species extinctions. In large part, this is because we are unsure of the actual number of Earth's species.

According to conservative estimates, Earth's biosphere contains between 7 million and 10 million species. Some groups of species have received the lion's share of attention from naturalists; consequently, we may have identified most of the living species belonging to these groups. Among them are birds (about 9,000 species), mammals (about 4,500 species), butterflies (about 19,000 species), and plants (about 250,000 species). On the other hand, many other groups of organisms are little known, so estimates of the numbers of species making up the groups are much less certain. For example, we have named only about 70,000 species of fungi, although as many as 1.5 million species may actually exist. Considering the great economic importance of the group—decomposers, agents of disease, agents for producing bread and alcoholic beverages, etc.—our ignorance of them is a serious problem. No one can make an accurate estimate of the number of species of bacteria, but the 4,800 named species surely represent only a tiny fraction of the actual total.

Overall, we have named only 1.5 million species of organisms (Figure 1). Approximately two-thirds of these are from temperate regions, where most of the world's scientists live. We have named fewer than half a million of the species living in the tropics, which may contain as many as 6 to 8 million species. In other

Figure 2: Passenger pigeon (*Ectopistes migratorius*), United States, extinct 1914.

words, there are probably fifteen to twenty unnamed tropical species for every named one, indicating a very important area for study.

Worldwide, species are becoming extinct 1,000 to 10,000 times faster than before our ancestors first appeared on Earth. According to the fossil record, the average life span of a species is about 4 million years. This means that just before humans appeared, the chance of an average species becoming extinct during any given year was 1 ÷ 4,000,000, or 0.00000025. If we estimate that Earth then contained 10 million species, the extinction rate would have been 10,000,000 x 0.00000025, or 2.5 species a year. Biologists call this the "background" or "natural" rate of extinction; it is the presumed rate without humans. Estimates show that over the next few decades, we could lose about 50,000 species per year, a rate 20,000 times the background rate. By the year 2100, perhaps two-thirds of Earth's current species will have disappeared or be on the way to extinction. Their loss will greatly limit human prospects for the future in ways that we can only dimly perceive.

Giving a precise estimate of the actual number of species being lost in the rain forests and other habitats is difficult. However, we could make reasonable estimates if we only knew the actual number of species in these environments. Ecologists have observed that an area of ten square miles contains twice as many species, on average, as does an area of one square mile. This suggests that reducing a particular area to a tenth of its original size will cause half its species to become extinct. Since 1945, our cutting and burning have decreased the forested area of Earth by one-third. Perhaps as much as nine-tenths of Earth's forested area will have disappeared by the year 2030. This would ultimately lead to the extinction of half the species now living in Earth's forests. How many species this would

be is uncertain because we lack accurate estimates of the number of species now living in these environments.

When we cut down a rain forest in Papua New Guinea or Ghana, we lose an entire forest community in which probably ninety-five percent of the species are unknown. Also lost is the possibility of learning more about the relationships among the many forms of life that live there together; this aspect of ecology is not well understood by scientists.

As the world's human population continues to grow, along with levels of consumption and rising expectations for a higher standard of living, so will the need for land and resources. This will inevitably lead to more extinction of species. Perhaps 2 million species, each the product of billions of years of evolution, could become extinct during our lifetimes. The biodiversity we are losing is the original source of nearly all sustainable productivity: our food, medicines, fiber, building materials, biomass for solar energy, and doubtless a host of potential new products we have not yet discovered. Each yet-to-be-discovered species could provide us a variety of useful products and may reveal an importance to other species that we cannot now even suspect.

Our awareness of the ecological harm we are causing offers us an immense—and incredibly challenging—responsibility. We must act as if we were the custodians of Earth's organisms and ecosystems. We must commit ourselves to stop, and if possible even to reverse, the current trend of destruction as soon as possible. This can be accomplished only by accepting our individual responsibility to limit levels of consumption for ourselves and the areas where we live, becoming active politically in pushing toward sustainability, and collectively achieving a stable and supportable population level in a socially just world.

Amy Vedder: Profile

Left: Maidenhair fern *(Adiantum pedatum)*
Right: Fragile fern *(Cystopteris fragilis)*
Bottom right: Royal fern *(Osmunda regalis)*

62 63

As Director of the Living Landscapes Program for the Wildlife Conservation Society, Dr. Amy Vedder spends several months of the year traveling to Africa, Latin America, and other places developing projects to benefit animals, plants, and people. Vedder was first exposed to the wealth of biological diversity in Africa—and to the economic difficulties facing rural Africans —as a Peace Corps volunteer more than two decades ago. Several years later, she and her husband established the Mountain Gorilla Project in the Virunga Mountains of Rwanda. Today, tourists bring money into the community when they visit the gorillas, supporting the people more sustainably than when they hunted these great apes and turned forest into farmland. After returning to the U.S., she directed the Africa Program of the WCS for six years, and is now working in a program with a more global focus.

Vedder looks at the big picture, promoting both the protection of the wildest places and the careful management of ecological buffers— such as grazed areas of the African savanna that are crucial to the survival of wildlife, but can benefit local people at the same time. In many projects, she has discovered that, like us, indigenous people do not always know and appreciate the natural resources in their community. Economic improvement is critical for people who have a hard time feeding their family; nevertheless, Vedder says, "We should not cut short our expectations that people will recognize other benefits of protecting species—benefits that enhance spiritual, aesthetic, or ethical aspects of our lives."

Chemical Prospecting:
The New Natural History

Thomas Eisner

Hardly anyone admits to being a naturalist anymore. Natural history used to be the most respectable of professions, before biology fragmented into today's multiplicity of subdisciplines. Simply put, natural history is the exploration of nature—the search for novelty in the world of living things. This novelty can be new species, new behaviors, new ecological interactions, new functions and structures, new materials, or anything else still unknown about organisms. Many contemporary biologists got their start as naturalists and remain naturalists at heart. As

Thomas Eisner is Jacob Gould Schurman Professor of Chemical Ecology at Cornell University.

grown-ups, they yearn to roam through nature to "have a look," as when they were youngsters. Yet as professionals, they tend to hide their love for natural history, just as the professional biological establishment tends to belittle its importance. This is profoundly regrettable, for natural history has gained new significance.

Biochemistry and molecular biology have taught us how to link almost everything we observe about living organisms to specific chemicals and chemical processes that occur in them. Therefore, whenever we discover novel adaptations in nature, we know that the organisms displaying these adaptations contain previously unknown chemicals. Through various techniques, we can isolate these chemicals and determine their molecular structures. Often, we can even discover and determine the chemical structure of the genes responsible for producing these chemicals. Given that many of these "new" substances have important properties—as medicines, perhaps— exploring living nature can be highly profitable. Exploring nature is thus equivalent to chemical prospecting.

Such a program of exploration can also increase our appreciation of nature. Nature is not just a source of useful materials; it is also a source of useful knowledge. For example, we can use our knowledge of the chemical structure of a natural substance to synthesize it in the lab or factory, or even to synthesize chemically related substances with more useful properties. Viewing nature as a source of applicable knowledge could have an enormous impact on conservation. It could set new goals for natural history and, by refocusing the process of discovery itself, could redefine the role of the naturalist explorer.

Exploration can be immensely enjoyable for the naturalist and can lead to unexpected findings.

Moreover, discoveries that seem trivial at first glance can turn out to be valuable on reflection and further inquiry. Let me illustrate by example.

Oil-Eating Bacteria

Hemisphaerota cyanea is a small blue chrysomelid beetle commonly found on palmetto plants (*Sabal* spp.) in the southeastern United States. It feeds on palmetto fronds (large leaves) both as a larva and as an adult. Ordinarily, the adult beetle walks with a loose hold. However, when disturbed, its feet hold so tightly to its leaf that it effectively resists the prodding of such enemies as ants. Indeed, it clings with such tenacity that a force 200 times its weight may fail to pull it off (Figure 1a).

H. cyanea secures its hold by adhering to the surface rather than by anchoring itself (by "digging into" the surface). The "soles" of its feet bear a dense mat of bristles; each bristle ends in a pad wetted by oil (Figures. 1b–e). Preliminary analyses showed the oil to consist of a mixture of long-chain hydrocarbon molecules. Each leg has about 10,000 pads, for a total of about 60,000. The oil comes from tiny glands that open at the base of the bristles. During ordinary walking, the beetle treads lightly, touching the ground with only a small fraction of its pads. However, when disturbed, it presses its six soles down flatly, so that all its pads touch ground.

We were interested in this adherence mechanism because scientists knew little about how insects secure their foothold while walking. *H. cyanea* was not unique in relying on wetted bristles for adherence. Many other insects, including possibly all chrysomelid beetles, have bristle-bearing feet very much like those of *H. cyanea*. However, *H. cyanea* has unusually many bristles, which enables it to use them for defense as well as for walking.

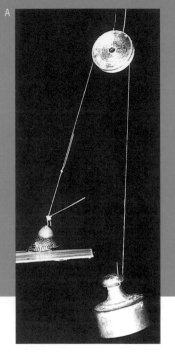

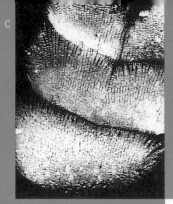

Figure 1: *Hemisphaerota Cyanea*

A   Beetle resisting load (2g).

B   Ventral view of beetle, showing the six broadly expanded foot soles.

C   Enlarged view of a foot sole, showing the bristles.

D   Enlarged view of bristle tips, showing the adherent pads.

E   Droplets of oil relinquished by adherent pads on contact with the substrate.

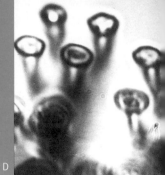

Our most interesting discovery about *H. cyanea* occurred by pure chance. The beetle has bacteria living on its feet that feed on the oil. We first noted these when we examined the oil droplets left by beetles in their wake when they walked on glass. We had some idea of the volatility of the oil and, on this basis, estimated how long it would take for the droplets to evaporate. Surprisingly, they vanished more quickly than predicted. As they disappeared, we noted the appearance at their margin of distinct rod-shaped bacterial bodies (Figure 2); in control experiments, these bacterial bodies did not appear at the edge of droplets of various other oils that we placed on the glass. The finding that these bacteria appeared to be "eating" the oil intrigued us; however, we lacked the appropriate expertise to culture and identify them. Our feeling is that isolating these microbes would be worthwhile. As oil eaters, they may have one or two useful and undiscovered enzymes hidden up their tiny sleeves.

New Biopolymers

Many animals, including amphibians, earthworms, and mollusks, have a body coating of slime. This coating provides protection against such small predators as ants, which may avoid sticky materials. Slugs use their slime to special advantage because they can coagulate it at specific sites where an enemy is attacking them. This is easy to show. If you gently poke a slug with a toothpick, nothing much will happen. However, if you simultaneously wiggle the toothpick, the slug will trigger the coagulation mechanism, which causes a rubbery blob to form around the tip of the probe. In my experience, this works with all slugs.

Needless to say, the coagulation mechanism, which appears to involve polymerization, provides for effective protection against small, biting enemies. Ants and carabid beetles, for instance, are truly muzzled when they bite into a slug. The chemical defense thwarts them the

E

moment they bear down with their mandibles (jaws); they back away with their mouthparts visibly encased in slime (Figures 3a–b).

How the coagulation occurs remains a mystery. The responses to electrical stimulation show that the coagulation begins within a fraction of a second. In at least some slugs, the coagulation coincides with the localized injection of crystalline material into the slime from specialized integumental (skin) cells. Moreover, we suspect that polymerization of proteins is involved. However, we know essentially nothing about the chemical details.

Slug slime also has other properties that deserve examination. In coagulated form, for instance, slug slime sticks with remarkable tenacity to human skin, including wet skin. Sticky materials abound in nature and could provide a fertile field for basic and applied research. Intriguing examples include the sticky spray of onychophorans (a phylum of "worms with legs") (Figures 3c–d), the gluey covering of dalcerid caterpillars (Figure 3e), the slimy coating of certain sawfly larvae (Figure 3f), and the sticky secretion from the hind parts of sow bugs (wood lice) and certain cockroaches.

Secrets of an Endangered Species

Dicerandra frutescens, a small herb with many branches, is a mint plant (family Lamiaceae) found only in central Florida (Figure 4a, see page 70). It is an inhabitant of the so-called Florida scrub, a highly interesting dry land ecosystem characterized by sandy ridges, shrubby plants, several vertebrates found there only, and a wealth of insects. *D. frutescens* has a distinct aroma that is so potent that we can sometimes detect the plant by its smell from a distance of meters downwind. Close examination of the plant revealed that it was virtually free of insect injury. Suspecting that the plant's odor was repellent to insects, we went on to find its cause. This turned out to be a terpenoid oil sealed airtight in tiny capsules on the leaves (Figures 4c–e, see page 70). (Terpenoids are a class of substances commonly present in plants.)

We thought that the capsules might function as chemical grenades. Insects would inevitably break some of them apart when biting into a leaf, causing the oil to spill out and then repel

Figure 2: *Hemisphaerota cyanea.* Portion of a beetle's oily footprint, photographed within hours of deposition (**A**); and days later (**B**). Note rod-shaped bacteria (arrows).

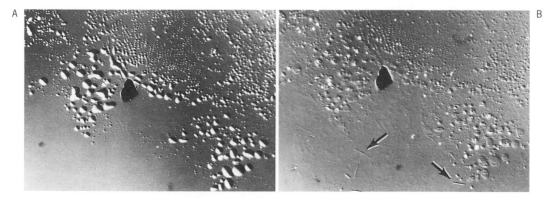

A   B

Figure 3:

A   Ant attack upon a slug. The ant (*Myrmecia* sp.) is with drawing, after having bitten into the slug and gotten its mouthparts gummed up with coagulated slime.

B   Minutes following the encounter, the ant is still attempting to rid itself of the slime.

C   An unidentified onychophoran (from Panama) discharging strands of its sticky secretion, in response to an "attack" with forceps.

D   The onychophoran secretion quickly assumes a rubbery consistency on exposure to air and becomes powerfully adhesive, even to human skin.

E   A dalcerid caterpillar (*Dalcerides ingenita*) showing its dorsal coating of rubbery gelatinous warts.

F   A sawfly larva (*Calicoa cerasi*) showing its glistening, sticky, integumental coating.

the insect by its odor. A simple experiment showed that the grenades really work. We quickly brought a freshly cut *D. frutescens* leaf near some ants attracted to a sugar source. This caused them to race away (Figure 4b, see page 70). In contrast, the ants tolerated intact leaves.

Chemical work subsequently showed the oil to contain twelve volatile terpenoid components. The principal component, whose chemical name is (+)-*trans*-pulegol, was a previously unknown natural substance. Finding a new insect repellent was exciting, especially because *D. frutescens* was an unusual plant. The species, first discovered in 1962, has a range of only a few hundred acres. It had already been placed on the Endangered Species List. If most of its acreage were not part of a protected site (the Archbold Biological Station), the plant would be in serious danger of being obliterated.

More remained to be found in this endangered species. Mycological work—mycology is the study of fungi—showed *D. frutescens* to contain more than twenty endosymbiotic fungi. (These are fungi that live inside the body of a host organism in a relationship that is mutually beneficial to them and to the host.) Some of these fungi are still unidentified, and researchers have yet to screen any of them exhaustively for biologically active materials; nevertheless, they have already found that one of these fungi is the source of an antifungal toxin.

Conclusions

The 1.5 million species we have described to date represent only a fraction of the estimated 10 to 20 million believed to exist on Earth. We have yet to discover most of what living nature is hiding from us.

The question is: Who will make the discoveries and for what purpose? Will it be the molecular biologists alone? Molecular biologists are committed to explaining life processes using the laws governing chemical and physical phenomena. Consequently, their main interest in investigating living things is to discover "new" chemicals in living things and, perhaps, the "new" genes that control the production of these chemicals. I would argue that the naturalist has at least as much to contribute to the goal of discovering new substances. As I showed in the three examples above, field

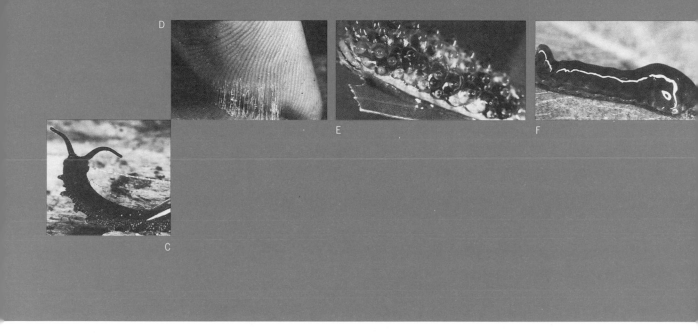

observation does not merely describe the appearance and behavior of living things; it also provides leads to previously undiscovered molecules and genes. The division between molecular biology and natural history is an artificial one. The practitioners of these disciplines would do best to work in concert.

Adding the mission of discovering "new" molecules does not fundamentally alter the discipline of natural history. What the naturalist has to offer will continue to have universal appeal and contribute, as it always has, to such established disciplines as ecology, evolution, behavior, and systematics (classification). However, because discovery in nature now has molecular implications, as well as vast potential for commercialization, the activity of the naturalist has new value. More so than any other scientist, the naturalist can list species in terms of their "chemical promise." Through observational skills alone, naturalists can categorize phenomena and point to those that might result from chemicals (and genes) of potential interest to medicine, agriculture, or material sciences. This makes the naturalist an extremely valuable member of the scientific

enterprise. Yet naturalists remain singularly unaware of their own worth, just as the commercial establishment is ignorant of what naturalists can provide.

We essentially ban natural history from universities and other institutions these days. Few academic courses teach students how to make discoveries outside the lab—as while exploring a tropical rain forest—or how to recognize the conventional and molecular implications of nonlaboratory observations. Naturalists will need to be trained worldwide and brought into the scientific mainstream. They are needed as explorers and as partners in applied science and industry. Most important, they are needed as spokesmen and spokeswomen for conservation. It is ultimately the explorer who is most aware of the value of what we could lose if the objects of exploration were to vanish. The example of the mint plant speaks for itself.

A

Figure 4: *Dicerandra fructescens*

A   Flower.

B   Ants, feeding at a sugary bait,
dispersing in response to approach
of a freshly transected leaf of the plant.

C   Portion of leaf, showing the
densely spaced, pearly oil capsules..

D   Enlarged view of an intact oil capsule.

E   Ruptured oil capsule. [Bars = 0.5 cm
(A), and 20 cm (D)]

B

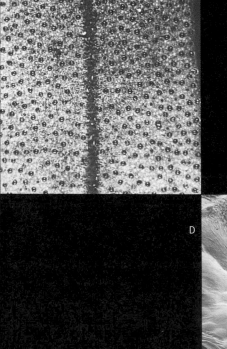

C

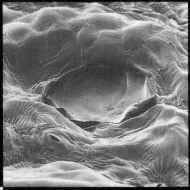

E

D

Dolores R. Santoliquido: Profile

That They May Survive: Ecuador, acrylic and color pencil, 28" x 34." All the animals and plants in this painting represent endangered species in the tropical rain forests of Ecuador.

When it comes to capturing the visual essence of a plant or animal species—what enables an observer to distinguish it from any other species—no photograph or written description can match a good line drawing provided by illustrator Dolores R. Santoliquido. Possessing a natural talent for drawing and trained as a sculptor, Santoliquido has a passion for the outdoors that helped propel her into a career as a natural science illustrator. Today she illustrates signage art for zoos, botanical gardens, and conservation parks—in addition to illustrating books and magazines concerned with natural science subject matter.

Illustrator, painter, and teacher, Santoliquido is committed to what she does. She says, "I consider myself fortunate to do the work I do.

Often I get to experience things and situations that someone restricted to the responsibilities of a conventional job cannot. There are many aspects of my work that are extremely rewarding: to work in the field, to visually document a species of plant or animal that is rare, or to see others learn from an illustration I completed for a conservation park or botanical garden. There are also aspects that are quite sad, when you return to a site where a particular species once flourished to find it diminished or no longer in existence there for one reason or another. It is also quite disturbing to realize that with every species that disappears there are numerous other species that are interdependent, and the disappearance of one may cause the disappearance of countless others."

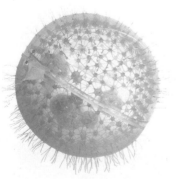

Volvox (*Volvox globator*)

Section Two: **Extinctions Past and Present**

Various beetles

Introduction Michael J. Novacek

Like individuals with a "life span" encompassing birth, survival, and death, species originate, endure over millions of years, and go extinct. Extinction is as much a part of the fabric of the evolution of life as is the origin and diversification of new kinds of organisms. The great preponderance of extinction took place as waves of turnover, where certain species were replaced by new ones. Under these conditions, the overall diversity of species may be maintained or may even be increased when new species branch off without going extinct themselves. In contrast, there are instances where huge sectors of life's diversity are obliterated, such as the mass extinction event at the end of the Cretaceous Period that counts dinosaurs as its most famous casualty. In this section, Niles Eldredge identifies five great extinction events that traumatized the biosphere over the past 450 million years. His central thesis is that we are now experiencing a sixth extinction event—one that is distinctively human-induced—of a scale and magnitude comparable to its predecessors. The human capacity to harness natural systems for agriculture and technology has led to domestication and degradation of the world's ecosystems by humans, as the importance of the local ecosystem becomes less and less relevant to the individual.

Humans have been instrumental in mass extinction prior to the current biodiversity crisis. Between about 40,000 and 1,500 years ago, the obliteration of many species of mammals, birds, and other vertebrates was coincident with the first appearance of humans in Australia, New Zealand, Madagascar, and many small islands. It has been popular to portray this correlation as mass extinction induced by overhunting by humans. Ross MacPhee and Preston Marx argue, however, that the most deleterious agents may have been infections from hyperdisease carried by humans and animals traveling with them. The authors outline molecular sampling for pathogens preserved in fossils that might be used to test their theory. It is clear that the impact of alien species introduced by humans disrupted ecosystems and promoted mass extinctions of birds and other native vertebrate species on many islands, a disturbing pattern vividly described by Helen James. On the more recent time scale, Ross MacPhee and Clare Flemming document the alarming loss of mammals since A.D. 1500 due to human activities. Hardest hit in the last 500 years have been small forms: rodents, bats, and insectivores. Surprisingly larger, more charismatic forms, such as rhinoceros, bear, and the big cats, have managed to survive during this same period of time.

Previous mass extinction episodes simply foreshadow the current situation. At local, regional, and global scales, the deterioration of habitats and decrease or loss of species is starkly evident. Several articles chronicle this trend for complex reef communities (see case study by Barbie Bischof), the fragile freshwater ecosystem of Lake Victoria (see case study by Melanie Stiassny), and the nesting habitats of songbirds in the forests of southern Illinois (see essay by Scott Robinson). Sometimes the forces in play are insidious but powerful. Theo Colburn, Dianne Dumanoski, and John Peterson Myers explain how the exquisite hormonal systems on which species depend for growth and reproduction are sabotaged by introduced chemicals. At the other end of the scale, Mark Bowen describes how ice cores, which record thousands of years of environmental history, provide data on climatic trends that may be disruptive to biodiversity. Although these studies leave many questions and predictions unanswered, they confirm the urgent nature of the current biodiversity crisis and its sobering similarity to mass extinction events of the past.

To explore past and present extinctions
I pose the following questions:

Are we in the middle of a sixth mass extinction?

Niles Eldredge, a Curator in the Division of Paleontology at the American Museum of Natural History and Chief Curator of the Museum's Hall of Biodiversity, thinks so, and also thinks that, unlike previous mass extinctions, this one is human-induced.

What are first-contact extinctions?

Ross D. E. Macphee, Curator of Vertebrate Zoology (Mammalogy) at the American Museum of Natural History, and **Preston A. Marx**, Senior Scientist at the Aaron Diamond AIDS Research Center in New York City and a Professor of Tropical Medicine and a Core Scientist at Tulane Regional Primate Research Center in New Orleans, argue that large-mammal extinctions, which occur only in certain places and in the presence of human habitation, are caused by hyperdisease carried by humans or the animals traveling with them.

What is the cause of island extinctions?

Helen F. James, Museum Specialist in the Department of Vertebrate Zoology, National Museum of Natural History, Smithsonian Institution, Washington, D.C., outlines the complex reasons for the extinction of island bird species both before and after the arrival of humans.

What can we learn about global warming from looking at ice cores extracted from tropical glaciers?

Mark Bowen, a freelance writer explains how the examination of oxygen isotopes in ice informs scientists of global temperature variations, both ancient and modern, and provides proof of global warming.

Why is there a decline nationwide in songbird populations?

Scott K. Robinson, Professor in and Head of the Department of Animal Biology at the University of Illinois at Urbana-Champaign and a Wildlife Ecologist at the Illinois Natural History Survey, explains the threats to songbirds caused by habitat fragmentation.

How do synthetic chemicals affect our lives?

Theo Colborn, Senior Program Scientist who directs the Wildlife and Contaminants Program at the World Wildlife Fund, **Dianne Dumanoski**, an author and veteran environmental journalist, and **John Peterson Myers**, Director of the W. Alton Jones Foundation, cite various studies that show the fatal effects of synthetic chemicals on animals and people and stress the need for caution and for further research to find alternatives.

Evolution, Extinction, and Humanity's Place in Nature

Niles Eldredge

Throughout the history of life on Earth, species have come and gone; those living today represent a very small fraction of the number of species that have once existed. More than ninety-nine percent of all species that have existed on Earth are extinct. Scientists estimate that as many as 4 billion species of plants and animals have existed at some time or other in the geologic past. Based on the known fossil record, most of these lived during the last 570 million years. (This period makes up the Phanerozoic Eon, which began with the first appearance of

Niles Eldredge is a Curator in the Division of Paleontology at the American Museum of Natural History and Chief Curator of the Museum's Hall of Biodiversity.

Late Cretaceous ammonite fossil from South Dakota. Ammonites went extinct at the end of the Cretaceous Period (AMNH 45282).

animals with skeletons.) However, this geologic eon has had severe episodes of mass extinction. During each of these mass extinctions, many species then living disappeared. Overall, the extinction of species has been almost as common as the origination of species.

What does extinction mean? Extinction is the disappearance of an entire species. Because it is difficult for paleontologists to document precisely species extinction, we often record mass extinction in terms of groups of species or families. It does not usually involve the sudden death of all its individual members. Rather, it is a consequence of the dynamics between rates of birth and death. When the death rate of the individuals in a species exceeds the birth rate for a sufficiently long period, the species becomes extinct.

Extinction and large-scale extinction are not new phenomena. Mass extinctions on a global scale have struck Earth's biotic systems five times since complex life first appeared on Earth just over one-half billion years ago (Figure 1). Geologists identify the major divisions of the geologic timescale with these mass extinctions, and with many other extinction events that were smaller and less widespread.

Five hundred seventy million years ago marked the beginning of the Cambrian Period. A great explosion of life and species diversity has occurred since then—as also have the five major extinctions.

One mass extinction occurred about 440 million years ago, near the end of the Ordovician Period. Up to that time, Earth's oceans had contained all its life. Some early corals and fish, which first appeared in the Ordovician Period, survived the extinction; however, many species of marine invertebrates did not. Near the end of the Devonian Period, about 370 million years ago, a second mass extinction occurred. It killed many species of fish and nearly seventy percent of the invertebrate sea creatures.

Perhaps the most devastating extinction occurred about 255 million years ago, at the end of the Permian Period and the beginning of the Triassic Period. Somewhere between eighty percent and ninety-six percent of living species perished. Because life on land had begun by this time, the Permian extinction affected both land dwellers and sea dwellers. The next major extinction—at the end of the Triassic Period, 200 million years ago—killed seventy-five percent of the sea-dwelling creatures and some

Figure 1: The Five Big Mass Extinctions: all taxa, Cambrian to Present.

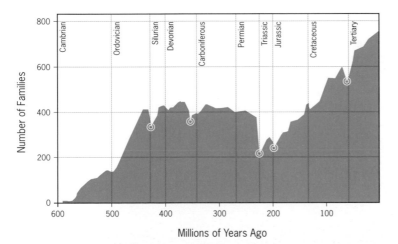

land species as well. However, the dinosaurs, crocodiles, and mammals survived.

All dinosaurs other than birds were the victims of the last great extinction, which occurred at the end of the Cretaceous Period, about 65 million years ago. This extinction killed about two-thirds of all species living at the time. Affected species included some kinds of plankton, corals and mollusks, all large marine reptiles, many other marine species, and most of the dinosaurs and their flying cousins, the winged pterodactyls. What survived? The turtles, crocodiles, small lizards, flowering plants and a few nonflowering plants, birds, and most species of mammals and fishes.

Scientists have many theories about what caused these five great mass extinctions, but no one knows exactly what happened. The extinction event at the end of the Permian Period appears to have resulted from a complex of environmental and physical causes. In contrast, the mass extinction at the end of the Cretaceous Period probably followed from the collision between Earth and one or more extraterrestrial objects, such as asteroids or comets.

Mass extinctions have profoundly altered the course of evolution of life on Earth. The more extensive the extinction, the greater the change in the species composition of the new ecosystems that reappear. (Establishment of these new ecosystems typically takes from 5 to 10 million years after the mass extinction event.) For example, before the mass extinction that marked the end of the Cretaceous Period, mammals had been small, scarce, and generally unimportant. However, once freed from domination by the dinosaurs, the mammals rapidly diversified. Mammals became the largest land animals and even evolved into

species adapted to a marine existence (whales, seals, etc.) and into species that could fly (bats). The informal term "The Age of Mammals," commonly applied to the current geologic era, suggests the great ecological importance mammals have achieved.

It is no longer possible to claim, with Charles Darwin and many modern evolutionary biologists, that extinction is a mere side effect of the evolutionary process. That is, that more highly evolved species outcompete and drive out competitively inferior, more primitive species. The evidence seems to show that extinction results primarily from changes—some slow, some rapid—in the physical environment. Life remains stable, with only minor "background" rates of extinction and evolution, until major ecosystem-wide disruption and extinction occur. This is the background or setting for many major episodes of evolutionary change, including the evolution of humans.

Today, the rich diversity of life is again threatened—not by a meteorite or huge volcanic eruptions, but by the actions of humans. We are currently in the midst of a sixth mass extinction. This one results from major changes of the planet's surface—caused not by natural climate change, but by our own species, *Homo sapiens*.

Over the past 4 million years, humans have evolved biologically; however, perhaps the most significant human evolution during this period has been cultural. (It should be kept in mind, however, that often a particular type of human cultural evolution depended on an earlier human biological evolution. For example, human toolmaking followed the biological evolution that resulted in hominids that walked on two feet and therefore had hands that were free to make and use tools. The development, much later, of human speech depended on the earlier biological evolution of the human "voice box," of a complex system of tongue muscles, and of a specialized region of the brain to control speech production.) The early phases of human evolution during this period were correlated with episodes of environmental change; the resulting environmental disruption and extinction of species led to spurts of evolution within both our own and many other lineages. The most spectacular of these episodes took place during a global cooling event that occurred about 2.5 million years ago. Then the wet woodlands of eastern and southern Africa changed to drier grasslands; this environmental change may have played a major role in the extinction of the early human-like primate *Australopithicus africanus* and in the evolution of two new species, *Paranthropus aethiopicus* and *Homo habilis* (the earliest known species of our genus).

The appearance of the species *Homo erectus* at the beginning of the Ice Age—1.65 million years ago—was the last time human biological evolution occurred simultaneously with, and perhaps resulted from, global climate change. From then on, cultural advances—for example, more sophisticated tools and the use of fire—played the major role in allowing *Homo erectus* to become better adapted to the environment. Humans then greatly expanded their range by moving north out of Africa and into Eurasia.

It is highly developed cultural objects—for example, tools, supplies, and equipment (including clothing)—that allow humans to live almost everywhere and anywhere. However, until about 10,000 years ago, humans still had a relationship to their local ecosystem that was typical of all other living things. Living in small groups of hunters and gatherers, they did not drastically alter their local ecosystems.

(However, as pointed out in the article by MacPhee and Marx, they may have indirectly caused the extinction of many species of large mammals.) In all likelihood, they had the same good sense of their local ecosystem—how it worked, how it affected them, and how they affected it—that anthropologists have observed in various hunting and gathering cultures.

The origin and development of agriculture approximately 10,000 years ago radically altered the relationship between humans and nature. (The development of agriculture closely followed the end of the Ice Ages. The resulting "drying" of the world's climate may have played a major role in stimulating humans to domesticate such animals as sheep, goats, pigs, and cattle, and such plants as wheat, legumes, potatoes, grapes, and dates.) Having a dependable fixed source of food made it possible for many people to live together in the same fixed location. As food production became more efficient, and as the human population increased, villages arose; eventually, some villages grew into cities—the hallmark of "civilization." Because of these cultural changes, human population has grown from an estimated 5 million to the current 6 billion in just 10,000 years. Demographers expect this number to double by the middle of the next century.

The very same process that led to the rise of human civilization led to a severe degradation of Earth's overall ecosystem. Whenever people changed a "natural ecosystem" to agricultural land, local biodiversity greatly diminished; what was once the habitat for a great diversity of species now supported a few species of agricultural crops and domesticated animals, and provided space for human habitation. As the human population has increased exponentially since then, much biodiversity has been lost through the domestication of many natural ecosystems.

In their increasing domestication of the world ecosystem, humans have increasingly come to "live outside" their local ecosystems. Of course, humans still live within the boundaries of their local ecosystem. However, once humans began to create food surpluses that they could store for use during droughts or other conditions that hampered agricultural productivity, they were much less "at the mercy of" their local ecosystems than previously. Once human societies began to trade with each other (so that they could import food when their local crops failed), they could truly "live outside" their local ecosystem—in the sense that they could receive their nutrients from outside their local ecosystem.

Moreover, as human societies became larger, people's economic and social roles became increasingly specialized. People in some social classes—for example, farmers and herders of domestic animals—still had to pay attention to their natural ecosystem to be successful. However, the well-being of an increasing proportion of people depended mainly on their success, as individuals and as members of a particular social class, in dealing with the other individuals and social classes in their society. In turn, the success of entire societies came to depend primarily on how well they could compete with, or defend themselves from, other societies. A consequence of this "increased socialization" of people was that they came increasingly to "live outside" their local ecosystem—in the sense that it became less and less important to them.

The increasing "separation" of people from their local ecosystems may have contributed to social policies that frequently resulted in severe degradation of their local ecosystems. The many cases of local ecological degradation have caused global ecological changes, which in turn could harm local ecosystems all over the world. For example, many localized instances of the burning of rain forest have increased the global atmospheric concentration of the greenhouse gas carbon dioxide. This may result in a global warming that could have harmful consequences to local climates everywhere.

Today, humans drain swamps and marshes to build houses. They burn rain forests to create cattle pastures. They cut down forest for fuel. They clear and transform land for farming. In addition, humans are manufacturing and using toxic chemicals, which end up in the environment. Cars, factories, power plants, and airplanes emit fumes that contain chemicals that are toxic to many living things. Farmers spray crops with chemicals to control pests. Many of these chemicals last long enough to contaminate the environment, poisoning creatures that eat them directly, or indirectly poisoning creatures that eat creatures that have these poisons in their bodies.

As we destroy habitats and spew chemicals into the air and water, plants and animals are disappearing rapidly. Extinction of both plants and animals continues today at an accelerated pace. Previous large-scale extinctions occurred over time spans of hundreds of years or longer. We humans are currently causing the sixth mass extinction Earth has experienced. Perhaps three species an hour, nearly 30,000 species a year, are currently being lost. The mass extinction caused by people may wipe out millions of species in a century or less.

Though we no longer live inside local ecosystems, we are part of the global system. The continued survival of our species depends very much on the health of this global system. We must stem the tide of the sixth mass extinction by curbing human population growth and our appetites for resources before it is too late—for the rest of the world's ecosystems and species—and for our own survival as well.

The island of St. Lucia.

St. Lucia Parrot
Recovery John Thomas

The St. Lucia parrot (*Amazona versicolor*) lives in rain forests on the small island of St. Lucia, in the Caribbean Sea. In the mid-1970s, the Jersey Wildlife Preservation Trust (based in Jersey, Channel Islands, United Kingdom) found only 100 parrots left in the wild. The major causes of the population decline were the destruction of its forest habitat, capture for the pet trade, and hunting.

In 1978, the St. Lucian Department of Forestry started an island-wide public awareness project to protect the parrot. This included education programs in schools and outreach to adults on conservation issues. These activities involved the participation of local businesses and citizen groups, which ensured support throughout the island's communities. In 1979, the parrot was officially designated as the national bird,

and the government established a parrot reserve and banned hunting.

In the early 1990s, the RARE Center for Tropical Conservation, based in Philadelphia, provided an additional boost to outreach efforts with its Promoting Protection Through Pride campaign. It also supported a traveling learning center called the Jacquot Express, after the French name for the St. Lucia parrot.

Since 1975, Jersey Trust has been doing population surveys for this species. In 1992, this work was expanded to include field research on foraging (food seeking) ecology, breeding and nesting cycles, and chick development conducted by its sister organization, Wildlife Preservation Trust International (WPTI), and by the Department of Forestry. With the knowledge gained from this research, reserve managers will better understand what environmental factors affect the survival of chicks and fledglings (offspring that have just grown their flight feathers). WPTI also supports the university-level instruction and field training of St. Lucian forestry personnel, which will give them the management skills to continue parrot conservation after international funding is withdrawn.

This protection and education project has had a dramatic effect: A 1996 census by WPTI and Forestry counted between 350 and 500 parrots. The recovery of this highly endangered parrot is a remarkable conservation success story, which shows how public support and the efforts of committed conservationists can contribute to the effective protection of biodiversity.

However, the St. Lucian parrot still faces serious threats. There is growing pressure to cut more forest to establish banana plantations and to allow hunting of other species in the parrot reserve. These threats emphasize the need for strong and strictly enforced laws if we are to protect wildlife populations in their natural habitat.

A field biologist measuring the wing length on a fledgling parrot.

Clare Flemming emerges from a cave in the West Indies after searching for remains of extinct mammals.

Clare Flemming: Profile

A Field Associate for the American Museum of Natural History's Department of Mammalogy, Clare Flemming spends much of her time climbing rocks, rappelling down cliffs, and crawling through tight spaces in caves. Why? "Because that's where the bones are, and bones are what I am looking for." An expert on a family of large extinct rodents that lived on Caribbean Islands, Flemming has found fragments of bones and teeth that have contributed greatly to the picture of what once lived on the island of Puerto Rico. "What we know today as rodents are a small fraction of what once existed. For instance, there are no native land mammals on the island of Puerto Rico. In fact, bats are the only native mammals living on the island. But several thousand years ago, very large rodents inhabited most of the island."

Flemming, who works with Ross MacPhee, Curator of Vertebrate Paleontology (Mammalogy) at the Museum, sees a clear connection between her research and biodiversity. "We don't know what is extinct and how much diversity once existed until we find evidence. This evidence is found as bone fragments and teeth that remain thousands of years after the animals have disappeared. Reconstructing a whole animal or group of animals from these little pieces is like doing a giant jigsaw puzzle about life."

The 40,000-Year Plague: Humans, Hyperdisease, and First-Contact Extinctions

Ross D. E. MacPhee and Preston A. Marx

Extinction is as much a part of the total pattern of life on this planet as is evolution. Without the winnowing effect of extinction, the production of new species by evolutionary forces would have overburdened the world's available habitats, which would have been a disaster for all living things. Nevertheless, although extinctions have occurred throughout the history of life on Earth,

Ross D. E. MacPhee is Curator of Vertebrate Zoology (Mammalogy) at the American Museum of Natural History. Preston A. Marx is Senior Scientist at the Aaron Diamond AIDS Research Center in New York City and a Professor of Tropical Medicine and a Core Scientist at Tulane Regional Primate Research Center of the Tulane Health Sciences Center in New Orleans.

the overwhelming majority of extinctions have occurred in clusters, some of great size. For example, during the past 500 million years, five mass extinctions involved estimated losses of up to ninety-five percent of marine invertebrate species. Extraterrestrial factors, such as asteroid impacts, may have caused some of these mass extinctions; however, scientists disagree about this, even in the famous case of the mass extinctions occurring at the end of the Cretaceous Period, when the dinosaurs (except birds) became extinct. Determining the factors that have caused these mass extinctions remains a great challenge.

Coordinated extinctions have also occurred on a much smaller scale, in which losses were significant but tended to affect only one or a few groups of organisms. By far, the most bizarre of these smaller extinction events occurred during the past 40,000 years; this time lies within the late Pleistocene and Recent (or Holocene), which together constitute the final part of the Quaternary Period. At different times on different continents and large islands, hundreds of land vertebrate species disappeared without being replaced by other species. The toll on species of large mammals—for example, the woolly mammoth, the saber-toothed cat, and the Irish giant deer—was especially high. The extinctions that took place in this interval have two unique aspects. First, they occurred at some locations but not at others. Losses were severe in Australia, New Guinea, North America, South America, New Zealand, Madagascar, and thousands of smaller islands, but they were negligible in the continents of Africa and Eurasia. Second, and perhaps more ominously, they all occurred after the emergence of anatomically modern human beings—*Homo sapiens*. What role, if any, did our human ancestors play in these extinction events?

According to one of the two major opposing views on this subject, our ancestors' role was merely that of the "innocent bystander." In this view, changes in climate that accompanied the end of the Ice Ages caused the extinctions. However, many extinctions occurred long after the conclusion of the last "Ice Age" and at locations that experienced no significant climate change.

According to the other major view, humans caused these extinctions, either directly by over-hunting, or indirectly by habitat destruction (perhaps by burning, in some areas). Compelling evidence supports the view that these extinctions would not have occurred without some role played by humans. Typically, large animal species became extinct shortly after humans reached specific locations for the first time. Scientists term these first-contact extinctions because they occurred shortly after the first-ever contact between humans and the victim species. Although the data are poor, it appears that the time from first contact to extinction may have ranged from one year on small islands to perhaps 1,500 years on large islands and continents.

How did humans cause these first-contact extinctions? One hypothesis—the "blitzkrieg" (lightning war) hypothesis—holds that when humans moved into new regions, they rapidly hunted some large animal species to extinction. However, several arguments tell against this hypothesis:

- Although many species of large mammals became extinct in North America at the end of the Pleistocene Epoch, researchers rarely find their remains in sites that were simultaneously occupied by humans. On the other hand, remains of now-extinct mammals are common in sites of various ages in Eurasia and Africa, where complete extinctions were rare during the late Quaternary.

Extinctions in Near Time: Pattern and Progress

Afroeurasia: No major episodes of extinction during the past 100,000 years, although some concentrated losses on offshore islands (e.g., Japan, 30,000 years ago) and in northern Asia (asterisk), affecting woolly mammoth, giant deer, and some other megafauna, in near-correlation with huge losses in New World around 11,000 years ago. Few losses in last 500 years.

Meganesia: Humans arrive 40–60,000 years ago; extinction episode follows at 50,000 years ago (including giant emu, *Genyornis*), but protracted phase of losses may also have occurred (up to 15,000 years ago?). Many losses in last 500 years.

Americas: Humans arrive 13,500 years ago or perhaps earlier; major extinction episode ends 10,500 years ago, with few subsequent losses.

Mediterranea: Humans arrive before 10–11,000 years ago on at least some islands, with extinctions of dwarf elephants and hippos around 10,500 years ago. Few losses in last 500 years.

Antillea: At least two large episodes, first around 5–7,000 years ago, shortly after Amerindian arrival, and again 500 years ago, correlated with appearance of Europeans. Many losses in last 500 years.

Madagascar: Humans arrive 2,000 years ago; major episode of extinction ends 500 years ago, with few losses thereafter.

Mascarenes (Ms): Humans arrive for the first time A.D. 1600; major episode of extinction (tortoises, dodo, solitaire, bats) ends around A.D. 1900.

New Zealand: Humans arrive 800 years ago; major extinction involving moa and other birds ends around A.D. 1500. Losses in last 500 years, especially among birds.

Commander Islands (C): Humans arrive A.D. 1741; Steller's sea cow extinct by A.D. 1768.

Wrangel Island (W): Humans arrive at least 3,100 years ago; latest woolly mammoth dates, 3,700 years ago (brief overlap probable but not yet demonstrated).

Galapagos Islands (G): Humans arrive A.D. 1535; many modern-era extinctions.

- On small islands, mammal species too small to have been the prey of human hunters suffer a high incidence of first-contact extinction.

- Extensive skeletal remains of mammoths occur in North America. However, most of the skeletons display no signs of butchery or indications that they were processed in any manner (for example, dismemberment). Mammoth "hunters" may in fact have been mammoth scavengers who opportunistically collected meat from animals already dead.

- After the period of first contact and initial substantial loss in affected areas, the extinction rate rapidly declined. For example, in the modern era (after A.D. 1500), only one species-level extinction in every eight has involved large animals; by contrast in the Americas 11,000 years ago, three of every four losses involved large animals. The first Native Americans were few and had limited tool kits. How could they have caused dozens of extinctions in a few centuries, whereas their descendants failed to cause a single extinction in the following 10,000 years?

To explain the phenomenon of first-contact extinctions, we have hypothesized that, under certain circumstances, disease can cause the complete extinction of species. As we show below, we hope to test this hypothesis experimentally.

Disease is rarely considered as a major immediate cause of extinction, yet under specific circumstances, it can cause catastrophic reductions in population size. We have recently argued that several fundamental features of the first-contact extinctions discussed here correspond to expected effects of massive population reductions caused by highly contagious, highly lethal diseases ("hyperdiseases").

One fundamental postulate of the hyperdisease hypothesis is that "new diseases"—ones that particular populations have encountered for the first time—can be extremely lethal; under

certain conditions, the result may be extremely rapid population crashes. Another postulate is that diseases capable of jumping to new species were able to travel with humans migrating to new lands. The migrating hosts of these diseases could have been humans themselves or organisms that depended on or interacted with humans and therefore moved with them (for example, certain species of insects and rats). A prime example of this may have been when humans first reached northern Asia and the Americas at the end of the Pleistocene Epoch. Many extinctions resulted from this first contact between humans and native animal species.

A particular pathogen would cause a "hyperdisease" only under the following conditions:

- It has a "reservoir species," a species that it lives in and/or reproduces in but does not kill or significantly harm. The pathogen emerges from the reservoir species repeatedly to infect other species.

- It represents a completely new challenge to host species not previously exposed to it.

- It typically kills its host before the host develops protective immune responses (antibodies, etc.).

- It spreads with a high mortality rate (greater than seventy-five percent) through all or most age groups of the host species.

- It infects many different species simultaneously without seriously affecting human populations. (As an example, rabies causes widespread, highly lethal epizootics in some animal species without causing epidemics in humans.)

Under hyperdisease conditions, population sizes of susceptible species would fall so low that the species would have been doomed to quick extinction. The hyperdisease did not have to infect, and kill, every individual. At very low population sizes, any number of random effects, including human hunting and environmental change, might potentially finish off the last survivors. Hyperdisease pathogens need not have been—and probably were not—completely new microorganisms or viruses. The lethality of a particular pathogen depends on whether a particular host can mount an effective immunological defense against it. Diseases that are harmless in one species may be acutely lethal in another. For example, the virus *Myxoma* has harmless effects in its natural host *Sylvilagus brasiliensis,* a South American cottontail rabbit. However, *Myxoma* had highly lethal effects (mortality rates of up to ninety-nine percent) when people introduced it into Australian populations of the European rabbit (*Oryctolagus cuniculus*).

Hyperdisease explains several key phenomena of first-contact extinctions just as well as, if not better than, its chief competitor—the blitzkrieg hypothesis:

- Extinction rates in affected areas consistently dropped off after an initial period of mass losses. Hyperdisease rapidly killed off the susceptible species, leaving species that were not vulnerable to the particular pathogen.

- Larger mammals are more susceptible to extinction than smaller ones are. Small mammals have higher reproductive rates and shorter gestations than do large mammals; therefore, their population could recover quickly after substantially decreasing.

- An extinction episode can pass with dramatic rapidity through an area whatever the local topography, plant cover, or the distribution of habitats favorable to human occupation. Infected members of the host species could spread the infection throughout their range, including portions not visited by humans.

- These extinctions did not occur in Africa or Eurasia (except for high-latitude areas). If humans had been in these areas long before 40,000 B.C., the era of first-contact extinctions in these regions would have been much earlier.

In principle, we can probably design a critical test that would prove that hyperdiseases have occurred in nature. At the very least, such a test should include the discovery of an identifiable pathogen in the preserved remains of several species that became extinct simultaneously. However, little work has been done in the field of "ancient" molecular pathology. One reason for this is that molecular investigators are normally interested only in the DNA (deoxyribonucleic acid) molecules that belong to the cells of their target organism; they are not usually interested in what is otherwise regarded as contaminant DNA. Another reason is that tests for DNA sequences are highly specific. This means that the odds of investigators identifying "foreign" DNA in a host organism are small unless they already know, or at least strongly suspect, the source of this foreign DNA. When classical techniques of pathology show the identity of the infecting microbe, then it is much simpler to detect the microbe's genetic material by conventional molecular techniques. (For example, tubercular lesions in pre-Columbian Indian mummy material strongly suggest that the tuberculosis bacillus was present in the New World prior to A.D. 1500. This finding was recently proved by finding molecular evidence of the bacillus' genetic code in a pre-Columbian burial in Ontario.) However, viruses, not bacteria, cause most acutely lethal diseases in mammals. Moreover, many extremely pathogenic viruses have RNA (ribonucleic acid) rather than DNA as their genetic material. This adds greatly to the complexity of detection because RNA molecules, which are single-stranded and therefore less stable than DNA molecules, typically do not survive nearly as long as DNA molecules do. Finally, at this time we do not know the identity of any hyperdisease pathogens.

In cooperation with Dr. A. Tikhonov (Zoological Institute, Russian Academy of Sciences) and Dr. S. Vartanyan (Wrangel Island State Reserve), we are beginning our search for a hyperdisease pathogen by sampling mammoth remains collected in eastern Siberia. Although people have found several other extinct species in Siberia, mammoth remains are the best reported and best collected. In addition, mammoths were geographically widespread, occupying northern Asia and North America at the end of the Pleistocene Epoch. This gives us the possibility of eventually being able to analyze the status of several different mammoth populations. We are proceeding on the assumption that bones will retain evidence of pathogens after death and that we will be able to characterize the genetic material of any pathogens we encounter by amplifying and sequencing it. Bones will be the crucial sampling source once we have identified candidate hyperdisease pathogens. With a definite target, we can use molecular probes specific for the genetic material of that pathogen.

From *Natural Change and Human Impact in Madagascar*, edited by Steven M. Goodman and Bruce D. Patterson; "The 40,000-Year Plague: Humans, Hyperdisease, and First-Contact Extinctions" by Ross D. E. MacPhee and Preston A. Marx; copyright © 1997 by the Smithsonian Institution. Used by permission of the publisher.

Prehistoric Extinctions and Ecological Changes on Oceanic Islands

Helen F. James

Introduction

Fossil evidence shows that many animals on the world's oceanic islands have become extinct during the Holocene Epoch, 10,000 years ago to the present. These extinctions were global in scope, occurring on islands in the Pacific, Atlantic, Arctic, and Indian oceans, and in the Mediterranean

Helen F. James is a Museum Specialist in the Department of Vertebrate Zoology at the National Museum of Natural History, Smithsonian Institution.

and Caribbean seas. Fossils of more than 200 species of extinct island birds have been found, for example. Many other species native to oceanic islands also became extinct, such as tortoises, lizards, bats, giant rodents, and dwarf hoofed mammals.

Paleontologists believe that continental extinctions during the Quaternary geologic period (1.64 million years ago to the present) were caused by such forces as climatic change, geophysical activity, and human impacts on the environment. Perhaps island extinctions during the same period had the same causes. During periods of maximum glaciation, a comparatively large amount of Earth's water was locked up in the polar ice caps. This caused Earth's sea level to be comparatively low, and this in turn caused shallow marine shelves that surround some islands to become exposed above sea level. Sometimes this resulted in land bridges that connected islands to nearby continents or to other islands, facilitating biological invasions by foreign species. Biological invasions can upset the ecological balance and cause extinctions. On the other hand, when glaciation was at a minimum, the sea level would be comparatively high. This would cause islands whose highest elevation was not significantly above sea level to be submerged partly or completely under water. This likewise could have caused some extinctions by reducing, or eliminating, the habitat of the islands' land animals. Catastrophic geophysical events such as volcanic eruptions and enormous avalanches are other potential causes of island extinctions. For example, volcanic eruptions can produce avalanches that cause much of the area of an island to sink beneath the sea.

However, the most widely held hypotheses to explain island extinctions during the Holocene Epoch assign a major role to human activity.

The fossil record shows that most island extinctions occurred after the first arrival of humans. Humans may have caused these extinctions directly by overharvesting plants and animals and by destroying habitat. Extinction could also have resulted from the arrival of competing species, predators, and even diseases that accompanied humans. The human-caused extinction of species native to islands is happening now and has occurred frequently in the historical past. For example, the extinction of the dodo, a flightless bird that once flourished on the island of Mauritius in the Indian Ocean, occurred after Europeans began visiting this uninhabited island in the sixteenth century.

To what extent can we attribute modern patterns of biodiversity on islands to natural factors? To what extent were these changes caused by humans? What caused the island extinctions, and how did these extinctions affect the functioning of the local ecosystem? In what follows, I review data about some aspects of these questions.

Diversity

Paleontologists have found evidence for Holocene vertebrate extinctions on most islands they have surveyed. The proportion of native vertebrates lost from individual islands varies. In the most extreme cases, some islands lost all their native land vertebrates (for example, Easter Island and Kahoolawe). New Zealand lost about thirty species of birds prehistorically. The Hawaiian Islands lost between thirty-five and fifty-seven species. Madagascar lost at least thirty species of mammals, birds, and tortoises.

The case of the Hawaiian Islands illustrates how prehistoric extinctions obscured our perception of how oceanic islands overall naturally developed their particular communities of animal species. When we consider the fossil record, we see that the number of bird species that successfully colonized the Hawaiian archipelago and diversified there (giving rise to a native group of related species) is much higher than that suggested by the bird species currently living there. Groups of related species that became entirely extinct in prehistoric times fall into two distinct categories. They are all either waterbirds (ibises, rails, geese, ducks), most of which adapted through evolution to land habitats after colonizing the islands, or raptorial birds (eagles, hawks, owls). No lineages of passerines ("perching birds," making up about sixty percent of today's bird species) became entirely extinct, and passerines suffered proportionately less extinction at the species level as well. In this way, extinction masked the success of waterbirds and raptors in colonizing the remote Hawaiian Islands and diversifying there; extinction also exaggerated the importance of passerines among the animal population.

On the Hawaiian islands of Maui and Oahu, roughly two species of birds have become extinct for every species now living. This high proportion of extinct species suggests that a prehistoric "faunal collapse" occurred. This affected not only the larger animal species that we conventionally regard as vulnerable, but also many smaller animal species that presumably would have been abundant and therefore less likely to become extinct.

The island record of fossil plants generally has not produced evidence of human-caused extinction on the same scale as for vertebrates. However, notable exceptions occur; for example, no native trees live on Easter Island, but lake cores contain pollen types representing about seven kinds of trees. These extinctions probably resulted from complete deforestation of the island by humans. Overall, prehistoric human-caused changes in island vegetation may have been extensive and may have had

important implications for the functioning of the local ecosystems. However, current evidence suggests that plants were more resistant to extinction than animals (at least vertebrates) in the same ecosystem.

Prehuman Extinctions

This distribution of island extinctions over long periods of time helps to convince paleontologists that human ecological disturbance caused most of them. Most available fossil evidence shows a cluster of extinctions in the human era and very few before then. For example, twenty-nine to thirty-four extinctions occurred in the Galápagos during the Holocene Epoch; all but three of these occurred in the five centuries since humans began visiting the islands.

Taking a step back in time, on the west coast of the South Island, New Zealand, the composition of bird species found in caves differs significantly between the last period of maximum glaciation (25,000 to 10,000 years ago in the Pleistocene Epoch) and the Holocene Epoch. Evidently, the climate change between the Pleistocene and the Holocene caused major regional shifts in the bird species distribution within New Zealand. However, as far as we know, no outright extinctions occurred. All the late Pleistocene fossils of land birds known so far from New Zealand are of species that survived until humans arrived, about 1,000 years ago. New Zealand bird species suffered far greater stress during the human era, when half the land species became extinct.

Taking another step back in time, on the island of 'Eua, Tonga, there are twenty-one species of birds in a fossil collection dating to between 60,000 and 80,000 years ago. Paleontologists traced these species' fate by checking for their fossil remains in deposits of human artifacts on 'Eua; these remains are less than 3,000 years old. These deposits contained remains of all but five of the twenty-one bird species, suggesting that at least three-fourths of the Pleistocene animal species survived until the human era. Eleven of these species (half the Pleistocene animal species) disappeared from 'Eua during the human era.

A deeper Pleistocene record is available from the Hawaiian island of Oahu. There, paleontologists recorded seventeen land-bird species in a preliminary study of Pleistocene fossils from Ulupau Head; the age of these fossils is more than 120,000 years. Between then and the present, global climate passed through a complete cycle of glaciation and deglaciation, a potential cause of change in animal species. However, in this case, very little change in the animal species occurred in prehuman times. All but perhaps one or two (twelve percent or less) of the species from Ulupau Head survived through the time of most-pronounced climate change. These species were still living in the mid- to late Holocene; tellingly, thirteen (seventy-six percent) of the Pleistocene survivors became extinct during the past few thousand years, when human impacts may have been a factor.

In summary, the fossil evidence argues for long-term stability of vertebrate communities on oceanic islands. Perhaps more surprising is the scarcity of evidence for climate-driven extinctions in the late Pleistocene, whether in tropical Hawaii or in temperate-to-subantarctic New Zealand. There are at least two possible explanations for this. One is simply that some island ecosystems resist climate-driven extinctions better than do some continental ecosystems. Another is that late Pleistocene climate change caused few extinctions, either on continents or on islands. Paleontologists still debate whether late Pleistocene extinctions on continents resulted mainly from the spread of archaic people or from climate change.

Prehuman Biological Invasions

Humans can cause disturbance on oceanic islands directly through their own activities; they also frequently help other alien species to invade, sometimes with very harmful results. However, we know that island ecosystems have been invaded and harmed by alien species often in the past and presumably suffered adverse consequences even before the appearance of humans. To put human-assisted invasions in perspective, we need to know how island ecosystems have responded to invasions when humans were absent.

For example, several lines of evidence suggest that the subfossil sea eagle of the Hawaiian Islands is a recent, but still prehuman, invader. First, the Hawaiian eagle does not show any bodily adaptations to the island environment. Paleontologists cannot distinguish its skeleton from that of either the white-tailed sea eagle, which lives in Eurasia north of the Himalayas, or the bald eagle, which lives in North America north of Mexico. Second, radiocarbon dating of eagle bones found in Puu Makua Cave on Maui showed that the bones are between 3,389 and 3,689 years old. This shows that the eagle was in the Hawaiian Islands at least 1,300 years before human arrival. Remains of the eagle were not found in deposits dated at greater than 120,000 years ago, a time at which most groups of Hawaiian birds were already present. The subsequent arrival of an eagle would have introduced a new upper level to the food chains of Hawaiian ecosystems, where flightless birds may have evolved without large predators. If ecological adjustments occurred in response to predation by the eagles, there is no evidence that they included rapid extinctions. Radio-carbon dating shows that the extinction of Hawaiian flightless birds took place centuries after humans arrived and at least 2,000 years after the arrival of the eagle.

Human-Caused Changes to Vertebrate Feeding Guilds

From giant owls in Cuba, to giant eagles in Crete and New Zealand, to foxes in Sardinia and Corsica, predators at the top of the local food chain have been particularly vulnerable to prehistoric extinction. Large, native herbivores were vulnerable as well. Moas, swans, and geese are gone from New Zealand; elephant birds, giant tortoises, pygmy hippos, and giant lemurs from Madagascar; ground sloths and monkeys from the Greater Antilles. Many, but not all, of these extinctions occurred after human arrival.

The Hawaiian Islands help illustrate how prehistoric extinctions altered the structure of vertebrate feeding guilds. (A guild is a group of species that play similar ecological roles within the same habitat.) Based on a very rough division of the resident bird population into guilds, extinction clearly affected some groups more than others. All but one species of native predators became extinct prehistorically. The one survivor (the Hawaiian hawk) became restricted in distribution to the largest island, so that most regions of the islands lost all native predators. To some extent, native raptors were replaced by owls that arrived after human settlement; however, the new arrivals tended to feed on rodents, while their extinct counterparts specialized in eating the native birds. Similarly, all land herbivores became extinct except the Hawaiian goose, which survived only on the largest island. Terrestrial omnivores (flightless rails and ibises) had been an important component of Hawaiian ecosystems, but they too all but disappeared prehistorically.

Most of the species of vertebrates that survived the Hawaiian extinction event are passerine birds. Of sixty-seven passerine species among the Holocene animals of the main islands, twenty-five (thirty-seven percent) became extinct prehistorically, and another twelve (twenty-two percent) have become extinct since European

contact. The proportions of extinctions among the insect-eating passerines and the fruit-eating or omnivorous passerines were the same as for passerines generally. In contrast, seed-eating passerines have suffered grievously from extinction, beginning in the prehistoric phase of extinctions and continuing historically; eventually they disappeared from all but one area of the main islands, high on the slopes of Mauna Kea. Meanwhile, nectar-eating passerines enjoyed the opposite fate. They began as the guild with the most species and remained so. Their extinctions were especially light during the prehistoric phase, when most of the seed eaters were disappearing. The changes that Hawaiian forest communities have undergone seem to favor certain nectar eaters, which now are the most abundant native birds in the forests. During the extinction event, predators disappeared, habitats shifted, and species richness plummeted. Therefore, it seems likely that some species, even some guilds, gained an advantage; perhaps they ended up being more widespread and abundant than they had been before the arrival of people.

Future Directions

More than any other factor, the Holocene fossil record of islands reflects the repercussions of one significant addition to the animal population, that of *Homo sapiens*. As humans became established on island after island, an extinction event followed. In effect, the experiment that can tell us how ecosystems respond to human-caused change, and particularly to reduced species richness, has already been set in motion. We have reproduced this experiment often on different islands, under varying conditions of climate, geography, ecosystem structure, human economy, and cultural evolution. The challenge is to use this source of information more effectively, to understand better what caused the extinctions and how ecosystems were affected. ✐

Reprinted from "Islands: Biological Diversity and Ecosystem Function", edited by Vitousek, P. M., L. L. Loope, and H. Adsersen (Ecological Studies 115), with permission from Springer-Verlag New York and Helen F. James.

* Although humans have a variety of eye colors, the overwhelming majority of all living mammals, and probably most extinct mammals as well, are or were brown-eyed.

Brown-Eyed, Milk-Giving… and Extinct: Losing Mammals Since A.D. 1500*

Ross D.E. MacPhee and Clare Flemming

Figure 1: *Xenothrix mcgregori* (jawbone, Jamaica 1996)

The Natural History of JAMAICA.   329

Readers of this book would by now probably agree that an imminent upsurge in extinctions, widely predicted, should be one of our leading concerns. But exactly how severe is the extinction problem?

As mammalogists, we felt that assessing the scope and pattern of species loss among mammals would be a feasible—and revealing—project. Mammals are now, and probably were always, relatively scarce. Fewer than 5,000 exist today. (In comparison, there are millions of species of insects.) Because mammals leave behind bones and teeth, we can identify species long after they have disappeared. Scientists probably know nearly all the mammal species that have lived during the past ten thousand years.

To discover very recent extinction patterns, we used scientific literature as well as records and chronicles by early explorers and travelers to compile a list of mammal species that have definitely disappeared since A.D. 1500. Our count: ninety mammal species. (See Figure 2 for the complete listing.) However, up to fifty percent of our listed species differed from those cited on similar lists by others.

Why were there so many discrepancies? One source of divergence was honest disagreement about classification. Some biologists see subspecies (varieties within a single species) where others see full species, and vice versa. A good example is provided by the quagga, the peculiar half-striped zebra that disappeared from southern Africa in the 1880s. Many scientists had considered it a full species. Recent molecular evidence, however, shows that the quagga is best regarded as a subspecies of Burchell's zebra, an animal still common in the region. Loss of subspecies, while regrettable, is different from the extinction of an entire species.

Incorrect extinction dates can explain other differences among lists. This was a particular problem with fossil species (those known only from their bones). Some, identified as modern extinctions, disappeared much earlier. For example, the extinct 300-pound Caribbean rodent *Amblyrhiza* is often listed as surviving into the sixteenth century; however, recent radiometric dating of some of its remains shows that this megarat actually died out 100,000 years ago. In contrast, many lists have excluded several species that survived into the modern era. In 1996, working in Jamaican caves with our colleague Donald McFarlane, we found remains of the extinct monkey *Xenothrix mcgregori* (Figure 1) mixed with the bones of European black rats; the black rat was introduced into the New World in Columbus' time. This circumstantial evidence—reinforced by an intriguing mention in a 1725 work by the renowned early naturalist Hans Sloane—indicates that *Xenothrix* survived into modern times. Remarkably, this is the only monkey species known to have died out in the last 500 years.

Our list reveals some patterns that defy the conventional wisdom on extinctions. According to the available evidence, for instance, major mammal losses have not occurred in the Amazon rain forest or the clear-cut forests of the United States or on Africa's Serengeti Plain. In fact, none of these regions—all on continental mainlands—has suffered a single, documented mammal extinction in the last 500 years! Australia is the only continent that ranks high on the mammalian extinction list. Continental Africa has lost only four species. The continental Americas, for all their biotic richness, show evidence of only one mammal species lost in the modern era. Globally, other vertebrate groups, such as birds, have suffered a much greater rate of loss in recent times.

Another puzzle: despite extraordinary levels of exploitation, only two marine mammal species have suffered extinction in the modern era: the Caribbean monk seal and the ten-ton Steller's sea cow. Remarkably, not one species of cetacean (whales and dolphins) has become extinct in the last 500 years, even though bowheads and other great whales were reduced to tiny populations in the early twentieth century. The gray whale came close to extinction when its entire Atlantic population was wiped out by the mid-1600s. Today, the species survives robustly only in the Pacific.

As can be seen on our map, the vast majority of modern-era mammal extinctions have occurred on islands. Other groups—including birds, reptiles, mollusks, and plants—have also suffered disproportionately on islands, many of which have lost all, or nearly all, their original biodiversity.

Ninety mammal extinctions in the past five centuries is our working figure. We would not be very surprised if, in the next few years, the number was revised upward to 110 or 115 confirmed losses. That's close to two percent of all mammal species on Earth. Does this sound like a little or a lot? Estimates of the "natural," or background, extinction rate for mammals run from one species each million years to one every 400 years. Let's choose the faster rate and do a quick calculation. Ninety species lost in five centuries represents a rate of one complete disappearance every five and one-half years—for a minimum 7,100 percent increase over the natural rate. Is this too high? Decide for yourself.

Excerpted from *Natural History*, April 1997.

96 | 97

Case Study

NORTH AMERICA

PACIFIC OCEAN

WEST INDIES

CANARY ISLANDS Volcano mouse*

CUBA
Cuban spiny rats (2 species)*
Cuban island shrews (4 species)*
Arredondo's solenodon*

HAITI/DOMINICAN REPUBLIC
Hutia (6 species)*
Quemi*
Hispaniolan spiny rats (2 species)*
Hispaniolan island shrews (3 species)*
Marcano's solenodon*

San Pedro Nolasco I. Pemberton's deer mouse **1931**

CAYMAN IS.
Cayman island shrews (2 species)*
Cayman coneys (two species)*
Cayman hutia*

PUERTO RICO Puerto Rican spiny rats (2 species)*

Maria Madre I. Nelson's rice rat **1897**

MEXICO Omilteme cottontail **1991**

BARBUDA Barbuda muskrat*

Little Swan I. Little Swan Island coney **1950s-60s**

CARIBBEAN Caribbean Monk seal **1950s**

MARTINIQUE Martinique muskrat **1902**
BARBADOS Barbados rice rat **1847-1890**
ST. VINCENT St. Vincent rice rat **1890s**

JAMAICA Jamaican rice rat **1877** Jamaican monkey*

ST. LUCIA St. Lucia muskrat **pre-1881**

GALÁPAGOS ISLANDS
San Salvador I. San Salvador rice rat **1965**
Santa Cruz I. Curio's large rice rat* Galápagos rice rat **1930s**
Isabela I. Large rice rat* Small rice rats (2 species)*

SOUTH AMERICA

ATLANTIC OCEAN

Figure 2:

Mammals: The Recently Departed

Hardest hit by extinctions in the last 500 years are rodents, bats and insectivores. Surprisingly, no rhinoceroses, bears, or cats of any sort have disappeared since A.D. 1500. despite their seemingly tenuous hold on survival at present.

*Species became extinct some time after A.D. 1500, precise date not known.

FALKLAND ISLANDS Falkland Islands dog **1876**

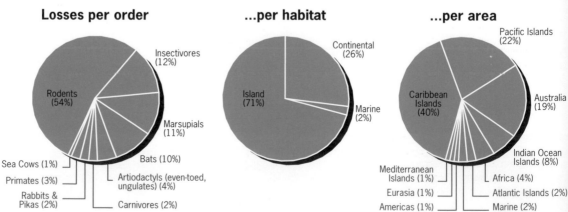

Losses per order
Insectivores (12%)
Rodents (54%)
Marsupials (11%)
Sea Cows (1%)
Primates (3%)
Rabbits & Pikas (2%)
Bats (10%)
Artiodactyls (even-toed, ungulates) (4%)
Carnivores (2%)

...per habitat
Continental (26%)
Island (71%)
Marine (2%)

...per area
Pacific Islands (22%)
Caribbean Islands (40%)
Australia (19%)
Indian Ocean Islands (8%)
Mediterranean Islands (1%)
Eurasia (1%)
Americas (1%)
Africa (4%)
Atlantic Islands (2%)
Marine (2%)

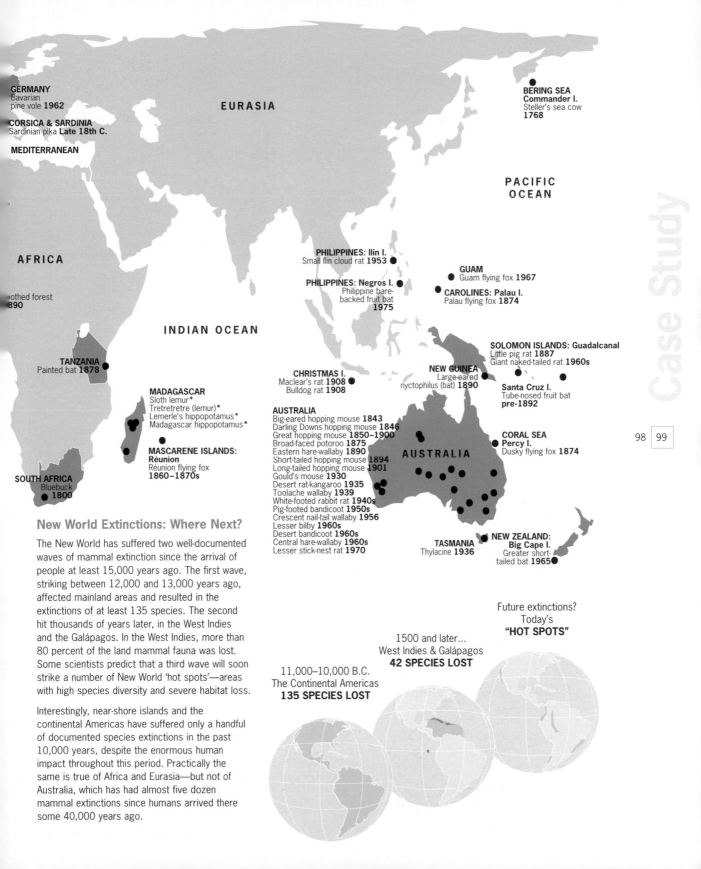

GERMANY
Bavarian
pine vole **1962**

CORSICA & SARDINIA
Sardinian pika **Late 18th C.**

MEDITERRANEAN

EURASIA

BERING SEA
Commander I.
Steller's sea cow
1768

**PACIFIC
OCEAN**

AFRICA

...othed forest
...890

INDIAN OCEAN

PHILIPPINES: Ilin I.
Small Ilin cloud rat **1953**

PHILIPPINES: Negros I.
Philippine bare-
backed fruit bat
1975

GUAM
Guam flying fox **1967**

CAROLINES: Palau I.
Palau flying fox **1874**

TANZANIA
Painted bat **1878**

MADAGASCAR
Sloth lemur*
Tretretretre (lemur)*
Lemerle's hippopotamus*
Madagascar hippopotamus*

MASCARENE ISLANDS:
Réunion
Réunion flying fox
1860–1870s

SOUTH AFRICA
Bluebuck
1800

CHRISTMAS I.
Maclear's rat **1908**
Bulldog rat **1908**

NEW GUINEA
Large-eared
nyctophilus (bat) **1890**

SOLOMON ISLANDS: Guadalcanal
Little pig rat **1887**
Giant naked-tailed rat **1960s**

Santa Cruz I.
Tube-nosed fruit bat
pre-1892

CORAL SEA
Percy I.
Dusky flying fox **1874**

AUSTRALIA
Big-eared hopping mouse **1843**
Darling Downs hopping mouse **1846**
Great hopping mouse **1850–1900**
Broad-faced potoroo **1875**
Eastern hare-wallaby **1890**
Short-tailed hopping mouse **1894**
Long-tailed hopping mouse **1901**
Gould's mouse **1930**
Desert rat-kangaroo **1935**
Toolache wallaby **1939**
White-footed rabbit rat **1940s**
Pig-footed bandicoot **1950s**
Crescent nail-tail wallaby **1956**
Lesser bilby **1960s**
Desert bandicoot **1960s**
Central hare-wallaby **1960s**
Lesser stick-nest rat **1970**

AUSTRALIA

TASMANIA
Thylacine **1936**

NEW ZEALAND:
Big Cape I.
Greater short-
tailed bat **1965**

98 99

New World Extinctions: Where Next?

The New World has suffered two well-documented
waves of mammal extinction since the arrival of
people at least 15,000 years ago. The first wave,
striking between 12,000 and 13,000 years ago,
affected mainland areas and resulted in the
extinctions of at least 135 species. The second
hit thousands of years later, in the West Indies
and the Galápagos. In the West Indies, more than
80 percent of the land mammal fauna was lost.
Some scientists predict that a third wave will soon
strike a number of New World 'hot spots'—areas
with high species diversity and severe habitat loss.

Interestingly, near-shore islands and the
continental Americas have suffered only a handful
of documented species extinctions in the past
10,000 years, despite the enormous human
impact throughout this period. Practically the
same is true of Africa and Eurasia—but not of
Australia, which has had almost five dozen
mammal extinctions since humans arrived there
some 40,000 years ago.

Future extinctions?
Today's
"HOT SPOTS"

1500 and later...
West Indies & Galápagos
42 SPECIES LOST

11,000–10,000 B.C.
The Continental Americas
135 SPECIES LOST

Global Warming, Loss of Habitat, and Pollution:

Introduction to "Thompson's Ice Corps," "Nest Gains, Nest Losses," and "Hormonal Sabotage"

Kefyn M. Catley

Many of the scientists in this book talk about the human-induced threats to Earth's biodiversity, such as loss of habitat, over-exploitation of species, the introduction of exotic species, pollution, and global warming. The three remaining essays in this section focus on the following specific threats: global warming, loss of habitat, and pollution. To help better understand these

Kefyn M. Catley is an Assistant Professor of Science Education at Rutgers, The State University of New Jersey.

essays, let us examine each of these causes of biodiversity loss in some detail.

Global Warming

Beginning with the Industrial Revolution of the late eighteenth century, the atmospheric concentration of carbon dioxide has increased dramatically, mainly because of the burning of such carbon-containing fossil fuels as coal, gasoline, and other petroleum-derived products. Also contributing to this increase has been the extensive burning of forests, especially in the tropics. (This burning not only adds carbon dioxide to the atmosphere, but also decreases the amount of global photosynthesis, thus decreasing the rate of carbon dioxide removal from the atmosphere.) During the past 100 years, the concentration of carbon dioxide in the atmosphere has increased from 290 parts per million to 360 parts per million, an increase of twenty-five percent; during the same period, the atmospheric concentration of methane has doubled (to about two parts per million). The world's climatologists have recently reached a consensus that the increased concentration of these and other greenhouse gases is causing global warming (Figure 1). If present trends continue, the atmospheric concentration of carbon dioxide could double by the end of the next century. This alone could cause Earth's average surface temperature to increase by an estimated 2°C. The ecological effects of such an increase could be devastating.

Acting analogously to the glass in a greenhouse, some atmospheric gases—water vapor, carbon dioxide, methane, nitrous oxide, and others—trap heat. They absorb infrared radiation from Earth's surface that otherwise would escape to outer space. This process is called the greenhouse effect. Without it, Earth's average surface temperature would be about 15°C colder than it now is. At this temperature, few of Earth's species could survive. Although Earth's

greenhouse effect is thus crucial to all its life-forms, any significant increase in the greenhouse effect could be extremely harmful to humans and biodiversity alike—by causing significant global warming. The increased melting of ice in Earth's polar regions would put entire islands and the low-lying coastal regions of continents under ocean water, resulting in property losses amounting to billions of dollars, loss of crop-growing areas, and saltwater contamination of freshwater aquifers. Harmful local effects would include increased rainfall and flooding in some locations and droughts (and possibly drought-triggered dust storms) in other locations. Tropical diseases would probably increase their range, infecting millions more people than they do today. As catastrophic as these effects may be, the harmful effects of global warming on biodiversity, and therefore ultimately on our own species, will be potentially even more devastating. The rise in sea temperatures may stop the upwelling of nutrients to the ocean's surface. This could cause ocean life to become extinct by eliminating the plant-like organisms at the base of marine food chains. Many marine and terrestrial species, already stressed by overexploitation and the loss and fragmentation of their habitat, would become extinct; the fast pace of the expected climate change will not give these species time to occupy new ranges. Increasing Earth's average temperature and increasing the availability of atmospheric carbon dioxide will affect plants in ways that scientists do not yet fully understand. However, one consequence will probably be a major change in plant biodiversity.

Preventing such consequences will involve significantly reducing our production of green-house gases. This, in turn, will require exploiting alternative sources of energy, decreasing consumption of energy (in part through technological improvements; in part by lifestyle changes), and preserving and restoring forests

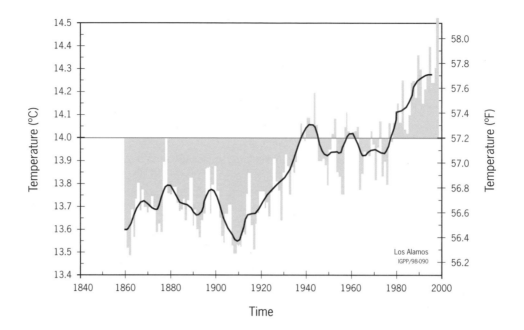

Figure 1: Measured global surface temperatones relative to the average since 1860. During the 30-year period between 1961 and 1990 there is a dramatic rise in temperature.

and other carbon dioxide absorbers. The world's political leaders are now struggling with the difficult challenge of enacting legislation aimed at reducing the production of greenhouse gases while simultaneously allowing sufficient economic growth to meet the needs of the world's ever-expanding population. Meanwhile, besides engaging in political activity aimed at the adoption of environment-friendly policies, we can take measures on the individual and community level to help reduce global warming.

The essay "Thompson's Ice Corps" is a great adventure story documenting the attempt to obtain a record of regional temperature changes during the past 14,000 years by collecting and analyzing ice cores from Andean glaciers. Information provided by this type of research can reveal the degree to which Earth's post–Ice Age climate has changed through "natural" processes.

Loss of Habitat

An increase in fragmentation results in an increase in edges. The edges of contiguous habitats such as forest are minimal. Breaking up habitats into smaller areas obviously increases edges. This phenomenon, known as the edge effect, has detrimental effects on the biota of the original habitat. Allowing more access often results in increases in invasive and pioneer species. Magnifying the effects of predation, parasitism, competition, and nonbiological environmental factors such as light and wind can have a profound effect on the ecosystem.

- Fragmentation can cause drastic changes in the structure of the biological community living in the original habitat, which may cause this local ecosystem to become unstable.

- Habitat fragmentation affects different species with different degrees of severity. Rare species, species with large ranges, and species with a low birth rate will suffer the worst consequences.

● Some effects of fragmentation may escape notice for tens or even hundreds of years. This is particularly so for changes affecting the structure of the biological community or affecting ecological processes.

In "Nest Gains, Nest Losses," Scott K. Robinson provides a vivid account of how habitat fragmentation affects various bird species in southern Illinois. These same effects have been noted in small woodland areas all over the United States, and on a global scale. Moreover, this fragmentation affects not only bird species but also whole forest communities of plants and animals. Most of these live in regions where the natural habitat is similarly fragmented. We can only imagine what our regions were like before they lost much of their biodiversity through this fragmentation.

As well as showing harmful consequences of habitat fragmentation, Robinson and his coworkers' research provides some basis for optimism by suggesting conservation approaches that may have a good chance of success. Overall, we need to maintain the integrity of what little natural habitat remains, minimize road construction, maintain and improve migration corridors between fragments of similar habitats, and manage existing fragments to decrease harmful effects, especially the edge effect. (The edge effect is a change in condition or in the types of species found where a habitat borders on another.)

Pollution

Almost every day we are faced with the results of pollution. Toxic heavy metals accumulate in fish and shellfish; nuclear plants continue to release airborne nuclear contaminants; beaches close because of sewage contamination; and novel substances are increasingly linked to breast cancer and other cancers, to decreased fertility, and to birth defects. Hormonal residues in meat are increasing, and fish kills occur with

frightening regularity in our rivers and lakes. Add to this the effects of air and water pollution such as global warming (from carbon dioxide and other greenhouse gases), eutrophication of lakes (from fertilizer runoff), acid rain (from nitrogen and sulfur oxides), and ozone depletion (from chlorofluorocarbon refrigerants)—and we begin to see our planet as a severely contaminated landfill.

Since 1962, when Rachel Carson's *Silent Spring* first pointed the finger at the dangers of pesticide dependency, their use has more than tripled. Each year approximately 2.5 million metric tons of herbicides, fungicides, and insecticides—50,000 different products in all—are applied to ecosystems throughout the planet. Some, such as organic phosphates, break down harmlessly. But many, such as DDT, PCBs, and other chlorinated hydrocarbons, cycle unchanged in ecosystems for many years. There they often become concentrated in animals approaching the top of the various food chains (biomagnification), with potentially catastrophic results. While the use of "persistent" insecticides has been restricted in the United States, their use continues in many developing countries. Apart from the effects of pesticides on human health, their impact on other nontarget organisms, especially beneficial insects, spiders, and other arthropods, are of serious concern. From an immediate human perspective, some synthetic chemicals are known to harm human health, others are suspect, and the majority we simply don't know much about.

Theo Colborn, Dianne Dumanoski, and John Peterson Myers's book *Our Stolen Future*, condensed sections of which are included here, presents a well-argued case that a variety of widely used synthetic chemicals that have found their way into the environment pose a severe risk to both human life and biodiversity because they "mimic" natural hormones. In studies of this nature, causal links between perceived agent and observed effects are extremely difficult to establish; consequently, not all scientists agree with all their findings. Nevertheless, the authors have done a timely service in focusing our attention on environmental pollution and its effects not only on the human condition but also on biodiversity.

The effects of all forms of pollution on biodiversity range from the documented case studies mentioned in the article to the disruption of entire food webs. Pollution's harmful effect on the biodiversity of terrestrial, freshwater, and marine ecosystems has been well documented. Europe's largest river, the Rhine, has seen a severe decline in species diversity during the past sixty years as a direct result of increasing pollution. Investigators have conclusively linked the disruption of the reproductive biology of many birds of prey and other birds to synthetic chemicals in the environment. They have also implicated pollution in cases of cancer in fish, mysterious diseases in marine mammals, and destruction of coral reefs by silt accumulation. Whole marine food webs have been disrupted by excess nutrients, typically fertilizer runoff

from land. The oxygen deficiency that has produced the so-called dead zone in the Gulf of Mexico for the past several years is a direct result of increased nutrient loading from the Mississippi River system. The excess nutrients stimulated an overgrowth of algae, which, after death, consumed oxygen dissolved in the Gulf waters as they decomposed.

These essays provide snapshots of ways scientists are studying some of the major threats to biodiversity. Only by scientific investigation will we begin to understand how to turn back the tide of human induced threats to biodiversity. There is still time to prevent many of the catastrophes that will inevitably result if these threats are left unchecked. The best way to ensure we leave our children a viable planet is to strive to produce scientifically literate and passionate environmental advocates.

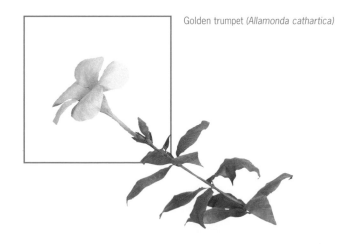

Golden trumpet (*Allamonda cathartica*)

Thompson's Ice Corps

Mark Bowen

The climbers I met in the bars of La Paz last summer, as I
adjusted to the altitude, yawned when I asked about the technical
difficulty of climbing 21,500-foot Nevado Sajama, Bolivia's
highest mountain. They called it a walk-up or a snow-slog,
although a long one. This was a good thing for the members
of the scientific expedition I was about to tag along on because
they had to get five tons of drilling equipment to the top.
Their mission? To extract ice cores from the summit glacier
and deliver them, frozen, to laboratories in Columbus, Ohio.

Mark Bowen is a freelance writer, photographer, and designer of medical instruments.

The leader of this expedition of more than fifty people—including scientists, technicians, mountaineers, and porters—was Lonnie Thompson, a paleoclimatologist from the Byrd Polar Research Center at Ohio State University. He hoped to obtain a record of changes in regional climate extending into the last glacial episode (the Ice Age), which ended about 14,000 years ago. To acquire these icy archives, Thompson and five others would live and work on Sajama's summit—roughly 4,000 feet higher than the base camp for Mt. Everest's summit—for twenty-eight days, a feat of high-altitude endurance that has been exceeded only twice in mountaineering history.

Sajama is a massive volcanic dome that sits alone near the western edge of the Altiplano, the world's second-highest plain. I first saw it at night: an enormous black hump under a bright stroke of the Milky Way, in an icy, star-filled sky. By this time, the summit crew had already drilled 100 meters of the first core—more than one-third of their goal—and arrangements were being made for porters to haul the ice down.

Five days later, after I had become acclimated to the 14,000-foot elevation of the base camp, I accompanied Todd Sowers, a geoscientist from Pennsylvania State University, on a round-trip to the high camp at 18,700 feet, the launching pad for the summit. Sowers described a few of the chemical assays that would be used to read the story in the ice. He would be responsible for analyzing the gas bubbles, while the analysis of the ice itself, and the dust and microbes embedded in it, would be carried out primarily in Thompson's labs.

One of the more useful markers in this branch of paleoclimatology is the isotopic composition of oxygen. In oxygen's predominant isotope, ^{16}O, the nucleus of the atom consists of eight protons and eight neutrons. The next most prevalent isotope, ^{18}O, has two extra neutrons. Water molecules containing ^{18}O have a stronger tendency to be in the liquid state or solid state (compared to the gaseous state) than do those containing ^{16}O. Quantitatively, the relatively stronger tendency for water molecules containing ^{18}O (compared to water molecules containing ^{16}O) to condense from the vapor state increases as the condensation temperature decreases.

As air, laden with water vapor, rises from the Atlantic and moves west toward the Andes, it encounters colder conditions the higher it goes. The portion of the vapor that condenses to form clouds—which, over the Andes, are typically composed of snow crystals—depends on temperature and pressure. If the crystals get big enough, they fall as snow. As previously mentioned, the $^{18}O/^{16}O$ ratio in the snow crystals depends on the temperature where they formed (through condensation of atmospheric water vapor). So the ratio found in Sajama's snow is a measure of the temperature of the air above it at the time the snow fell.

In addition, in oxygen from ancient atmospheres found in gas bubbles trapped in Sajama's ice, the $^{18}O/^{16}O$ ratio corresponds to the size of Earth's continental ice sheets when Sajama's ice was formed. Continental ice sheets grow by accumulating snowfall. When water vapor evaporates from the ocean, it forms atmospheric water vapor that is initially rich in ^{16}O relative to ^{18}O (leaving behind ocean water that is enriched in ^{18}O relative to ^{16}O). As this water-vapor-rich air mass moves toward the poles, decreasing temperatures cause water vapor in the air to condense to rain and snow.

The condensed rain and snow are rich in ^{18}O and poor in ^{16}O, and the water vapor remaining in the air is therefore rich in ^{16}O poor in ^{18}O (because the total quantity of these two isotopes on Earth remains constant). By the time the air mass gets near to either pole, the snow that condenses out ends up being enriched in ^{16}O relative to the rain and snow that condensed out closer to the source of the water vapor. This ^{16}O-rich (or "light") snow then accumulates to form continental ice masses. Meanwhile, the ^{18}O-rich rain that condensed from the air mass on its way to either pole eventually makes its way back to the oceans, further enriching the oceans in ^{18}O. So while the continental ice sheets grow by the accumulation of ^{16}O-rich snow in the higher latitudes, the oceans become progressively enriched in ^{18}O. Unlike the oxygen in the water molecules in Earth's ice, the oxygen in the oceans' water molecules is constantly being exchanged, by a variety of chemical processes, with oxygen in Earth's atmosphere. Consequently the ratio of ^{18}O to ^{16}O in Earth's atmosphere corresponds to the ratio of these two isotopes in the ocean—and thus, for reasons discussed above, corresponds to the total amount of ice that has collected in Earth's continental ice sheets. This is why the $^{18}O/^{16}O$ ratio in air bubbles trapped in particular layers of Sajama's glacier indicates the total size of Earth's continental ice sheets at the time these bubbles were trapped.

Because the extent of Earth's glaciation has varied over time, the $^{18}O/^{16}O$ ratio in the air bubbles trapped in particular layers of ice sheets can indicate the age of these layers—through comparison of the oxygen isotope ratio in these layers with the isotope ratios in other layers (different ice sheets) of known age. (In gas bubbles trapped 20,000 years ago—during the last glacial episode—the ratio was 1.3 ^{18}O atoms per thousand ^{16}O atoms; this ratio is higher that that of today's atmosphere.) Sowers hopes to obtain precise dates for the deep portions of Sajama's cores by comparing the ratios he measures in them against standard curves from Greenland and Antarctic cores. Greenland ice cores can be dated very accurately because their annual layers can be distinguished individually as far back as 45,000 years. He will also measure atmospheric carbon dioxide and methane, the second and third most important greenhouse gases (water vapor is the first). In polar cores, researchers have found the levels of these gases to be high during warm periods and low during cold periods. About 12,000 years ago, a cold event called the Younger Dryas caused the amount of glaciation first to increase and then to decrease. Sowers expects to find a distinct dip in methane levels, for instance, in the ice from that period.

Tropical glaciers—planetary archives that provide the raw data on which Thompson's life's work is based—are fast disappearing. Thompson believes that most glaciers, including Sajama's, will be gone within fifty years, along with their life-giving water and the power it generates. Almost all climate scientists see the melting of such glaciers as an early sign of global warming. He describes returning in 1991 to a glacier on Quelccaya, in southern Peru, where he had obtained a fine 1,500-year-old ice core in 1983. In the intervening years, meltwater from the glacial surface had seeped into the ice and obliterated the record. "The zero degree isotherm (the lowest altitude at which ice will remain frozen year-round) is rising at 4.5 meters per year," he insists. "Three glaciers in Venezuela have disappeared since 1972. Forty percent of the glacier on Mount Kenya has disappeared since 1963. Every tropical glacier we have measured is retreating, and wherever we have sequential information, we can show that the rate of retreat is exponential."

To him these vanishing records are an important key to understanding global climate. "To be relevant, you must work in the tropics," he says. "They contain fifty percent of Earth's landmass and eighty percent of the population. Their heat drives global weather. If a subtle increase in carbon dioxide raises their temperature even slightly, more moisture will enter the atmosphere. The greenhouse effect of the vapor will cause worldwide temperatures to rise significantly."

Twenty years ago, Thompson's colleagues assured him it would be impossible to obtain ice cores from the Andes and the Tibetan Plateau, but the records he retrieved from both places have greatly advanced our understanding of global connections in Earth's climate system, the connection between El Niño and Asian monsoons in particular. Says Thompson, "A warmer climate may induce a permanent El Niño. Before 5,000 years ago, the people on the coasts of Ecuador and Peru ate tropical fish. Cold water appeared only recently. Since Earth was warmer then, that time may show the behavior of El Niño in a warmer climate. It won't be good for Australia and India, or for Ecuador and Peru."

El Niño is signaled by a weakening of the trade winds that normally drive warm surface water west from South America's Pacific Coast and pull a current of cold water up from the deep ocean to replace it. Fishermen along the coast of Peru, noticing warm seas near Christmastime, named the phenomenon El Niño after the Christ child. Since the cold-water fish they usually harvest disappear then, they spend the season repairing their nets. The warm water normally pushed to the western Pacific by the trade winds produces massive anvils of cumulus clouds near Indonesia. During El Niño, when the trade winds recede, the western Pacific and the Andes usually experience drought, while coastal Ecuador and Peru are hammered by rain. The

monsoon fails in India. Droughts occur in Africa and Australia. Storms pummel the coasts of California and Mexico. The 1982–83 El Niño, by most climatological measures the strongest of the twentieth century, killed nearly 2,000 people, left 600,000 homeless, and caused an estimated $16 billion of damage. Thompson and Bernard Francou intend to use a forty-meter core from Sajama to study El Niño, which may serve as a scaled-down model of the sort of abrupt climate change that might be induced by global warming.

Standing near his drill on the summit, with the wind-packed snow turning gold in the afternoon light, Thompson told me why he had chosen this spot. We were standing on the southern-most tropical glacier in South America, he explained. These cores would fill a gap between records he had previously obtained to the north, on Huascarán and Quelccaya, and to the south, on Antarctica's Dyer Plateau. They would add an important data point to a transect (sample strip) of the Americas in an international project called Past Global Changes Pole Equator Pole I and complement other cores he had obtained on the Tibetan Plateau, where the record displays long-distance correlation with El Niño–modulated South American weather.

He had put years of effort into choosing this site, and Sajama met his needs perfectly. One advantage is its great height. In the tropics, glacial-stage ice is found only at the deepest levels, frozen to bedrock. With the warming of the atmosphere, all but the highest glaciers now have rivers of meltwater flowing beneath them. The flow of Sajama's symmetrical ice dome originates at its summit, where it is coldest. The points where glaciers originate, such as mountain summits, are the best places to drill, because glacial flow is less likely to have fractured the strata of the ice there. In the High Andes, a new layer of snow is laid down each year during the wet southern summer. Dust is deposited on its surface during the dry winter. If the strata remain undisturbed, the snow of recent years is separated by dust bands, forming distinctly separate layers that look like tree rings. The weight of newer snow compresses earlier layers, so their thickness decreases with depth. Below roughly one hundred meters, individual layers become difficult to distinguish, but unusual events, such as volcanic eruptions that leave a layer of ash, may permit very precise dating.

Despite the drain of living at such heights, Thompson seemed tireless as he briefed me while sorting out problems with the drill and juggling expedition logistics over the radio. It seemed odd to be discussing global warming on benches carved of snow in a cave at the summit of an ice-capped mountain, but Thompson's concern was noticeable. "Our cores from eastern Tibet show that the last fifty years there have been the warmest in the last 12,000. On Huascarán and in some other places 10,000 years ago, if you look at the planet as a whole, this century seems to be the warmest since the last Ice Age."

Evidence of climate change in the ice-core record often coincides with dramatic dislocations in human activity. A century-long dust event, indicative of drought, is recorded in the ice from Huascarán dating from the third millennium B.C., a period also marked by collapsing civilizations and migrations in Europe, Asia, the Middle East, Africa, and South America. "Sometimes a difference of only a few degrees can cause the end of a civilization," Thompson observes, mentioning a dust event in A.D. 900, recorded in the Quelccaya cores, that seems to coincide with the abandonment of fields around Lake Titicaca by the pre–Inca Tiwanaku civilization.

Warmer weather is not the only indication of global warming. The extreme weather that has made headlines all through this decade may also be a sign. The frequent and extreme hurricanes in the Atlantic, for instance, or the severe rains and flooding in the Great Plains of the United States, or intensified El Niños.

Three weeks later, I visited Thompson in Columbus. Sajama's cores were stored safely down the hall. "The porters did well," he pronounced. "The ice core is in great shape. There's still fresh snow from the summit on the outside of the tubes, and the thermistors in some of the boxes never gave readings above 10°C. I expect these cores to cover 20,000-plus years, and the two 132-meter cores to allow us to cross-check the record." (By mid-November he had verified this: The bottom thirty meters of the deep cores exhibit a drop of 5.4 parts per 1,000 in the oxygen isotope ratio—similar to that in glacial-stage ice from Huascarán.)

He was vibrant with success. He looked fifteen to twenty years younger than he had on Sajama. Evidently, the only ill effect of his stay at the summit was the loss of his toenails to "frostnip," and they wouldn't be back soon. The following day he left for a site on the highest plain in the world: the Tibetan Plateau. Over thirty-five days in September and October, he and some other scientists obtained ice cores at the highest altitude ever: three cores drilled to bedrock, at 23,500 feet—breaking the record they had just set on Sajama.

Excerpted from *Natural History*, February 1998.

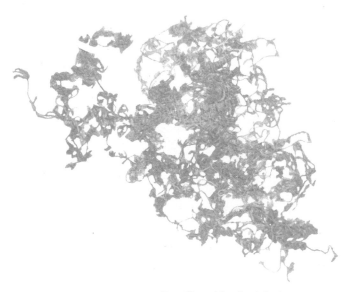

Moss *(Squamidium isocladum)*

Nest Gains, Nest Losses

Scott K. Robinson

More than sixty species of songbirds nest in the cypress swamps, the steep and oak-dominated hillsides, and the rich bottomland forests of southern Illinois. Year-round residents, such as woodpeckers, chickadees, and cardinals, are joined in summer by tanagers, warblers, flycatchers, and other migrants that arrive here to breed after making a long journey from their wintering grounds in Central and South America and the Caribbean.

Scott K. Robinson is a Wildlife Ecologist at the Illinois Natural History Survey and a Professor in and Head of the Department of Animal Biology at the University of Illinois at Urbana–Champaign.

Figure 1: A Red-eyed vireo *(Vireo olivaceus)* feeding a Brown-headed cowbird *(Molothrus ater)* chick and two Red-eyed vireo chicks. The biggest nestling is the Brown-headed cowbird.

At dawn on a moorning in early summer, two dozen fieldworkers from the Illinois Natural History Survey begin the day's tasks: Some count singing birds, others search for nests, and still others check the progress of nests found previously. At some census points—especially in the seasonally flooded bottomland forests—the standard five-minute period is too short for even experienced census takers to count all the singing birds.

Searching for nests is more difficult and requires great patience and concentration. At the end of the day, nest searchers gather to describe the nests they found and the cues that gave away their location. That day, they had found more than fifty new nests, in spite of the birds' best efforts to conceal them. For skilled searchers, work in these forests is a rewarding and productive way to spend a summer.

Less cheerful is the talk among the nest checkers, whose arduous job it is to examine the contents of nests periodically until the young fledge—that is, leave the nest and become independent—or until the eggs or young are found and eaten by predators.

Nest checkers wonder why birds breed at all in these forests. The data sheets for the nests monitored that day show that most contain at least one brown-headed cowbird egg. Cowbirds parasitize the nests of other birds, often dumping one of the host bird's eggs before laying one of their own for the host to raise. More than eighty percent of wood thrush, scarlet tanager, and summer tanager nests in these forests have been parasitized, and most contain two or three alien eggs. Although wood thrush eggs are pale blue and cowbird eggs are brown-speckled, thrushes fail to distinguish between them; instead, the thrush incubates the intruder's egg and tends and feeds the nestling as its own (see Figure 2).

Cowbirds parasitize the nests of more than 200 avian species. Generally, they choose birds that are smaller than themselves. Warblers, vireos, and some flycatchers, for example, have about half the weight of a cowbird. The thick-shelled cowbird eggs (small for the size of the female cowbird) develop faster and hatch earlier than the eggs of most hosts. The nestlings that emerge are usually larger than the host's young and eventually bigger than most of their surrogate parents. Their relentless demands not only tax the endurance of the smaller birds but also mean less food, care, and space for the rightful nestlings. Such parasitism can be devastating because many songbirds exploited by cowbirds fail to raise any young of their own. The larger the host, the more likely it is to be able to raise mixed broods of cowbirds and its own offspring; cowbirds seem to compensate for this by laying more eggs in nests of larger species, such as wood thrushes and tanagers.

Nest records also reveal a second problem, faced by all nesting birds. In any forest habitat, nests are lost to predators; a rate as high as

Figure 2: Brown-headed cowbird egg in nest of another bird.

fifty percent is not unusual. But in southern Illinois, losses to predation typically range from sixty to eighty-five percent, and most of our nest records quickly end up in the "defunct" file. When checkers find a prematurely empty nest, they try to piece together the evidence to determine which predator was responsible. Some nests have been torn apart, a sign of marauding raccoons. But more often the contents simply are gone, and the nest itself is undisturbed. In these cases, likely attackers include snakes, birds such as blue jays, or even small mammals. In southern Illinois, the most frequent culprit is the black rat snake.

What accounts for the discrepancy between the great number of songbirds counted by our census team and the great number of failed nests found by the checkers? This paradox has stimulated much of my research. Where do these birds come from if they are not adults returning to the forests where they hatched? Some of the populations are declining; others are not. Do a few species have tricks that enable them to cope with parasitism and predation better than we think?

Answers may come from research that stretches well beyond the borders of our study area in Illinois. At an informal meeting in 1990, a group of researchers working in neighboring states agreed that understanding local bird population trends requires a study of nesting over much of the Midwest. This group included Don Whitehead and his students at Indiana University, Frank Thompson of the U.S. Forest Service, John Faaborg and his students at the University of Missouri, and my group from the Illinois Natural History Survey and University of Illinois. We decided to conduct a coordinated study of cowbird parasitism and nest predation in forest tracts ranging from tiny woodlots in agricultural landscapes (ninety-five percent farmland) to the mostly adjoining forests of northern Wisconsin, the Missouri Ozarks, and south-central Indiana (all more than eighty percent forested).

The results were dramatic. Every bird species studied fared better in mostly forested areas, particularly in terms of exploitation by cowbirds. In extensive forests, cowbirds posed no serious threat to local songbird populations. On land largely given over to agriculture, however, cowbirds prevailed. Songbirds could find no forests far enough from farms and pastures, where cowbirds feed, to provide a cowbird-free refuge. Nowhere in Illinois have I found a forest tract that is more than four miles (the cowbird's daily commuting range) from a farm, pasture, or yard. As a result, cowbirds can reach any nest. Overall, parasitism levels were five to ten times higher in agricultural regions than in forested regions.

Similarly, predation is often greater in woods bordering yards, farmland, roads, and cuts for power lines. But even in the middle of the largest tracts of forests—those of more than 5,000 acres—parasitism and predation take a heavy toll. Fragmented forests—that is, wooded islands in a sea of cleared land—appear to be population "sinks" for songbirds. The annual number of young that fledge drops below the annual number of natural deaths, and the bird population is no longer self-sustaining. In contrast, more extensive forests probably serve as population "sources"; emigrants from these forests that settle in small woodlots reinforce and thus help rescue the threatened populations there, even when the woodlots are 100 miles from more extensive forests. In southern Illinois, the presence of some breeding songbirds may depend on the existence of large forests in the Ozarks.

Proving that local populations are in sinks requires detailed studies of particular species, and these studies have produced some surprises. Cheri Trine, a University of Illinois graduate student, observes individually marked wood thrushes in three of the largest forest tracts in Illinois. Wood thrushes can raise two and sometimes even three broods a season; they live longer than most songbirds; and they are large enough to raise cowbird young alongside their own offspring. So, at first glance, the wood thrush would seem to be able to weather the predation and parasitism in fragmented forests. But Trine found that between losses to cowbirds and to predators, each of her three wood thrush populations failed to fledge enough young to replace adults that died. For wood

Figure 3: Eastern wood-pewee (Contopus virens)

Figure 4: White-eyed vireo (Vireo griseus)

thrushes, even some of the larger fragments are sinks; the birds probably depend on extensive forests far from southern Illinois to keep up their numbers. The same may be true for red-eyed vireos, scarlet tanagers, summer tanagers, and hooded warblers. These are sobering results for anyone trying to conserve songbirds on Illinois farmland. None of our existing tracts of forest may be large enough for the species.

Unlike wood thrushes, Kentucky warblers are comparatively short-lived, raise just one brood per year, and rarely raise more than a single nestling of their own if they have been parasitized by a cowbird. These traits should make them extremely vulnerable to forest fragmentation. But according to studies conducted by University of Illinois graduate student Lonny Morse, the vulnerability of Kentucky warblers shows great variation within large forests. Populations of this species nesting deep within older forest stands—and far from farms—may be holding their own. To be self-sustaining, then, Kentucky warblers need cores of undisturbed forest within already large tracts.

Our regionwide study has enabled us to examine the needs of each species separately. And although some are in dire trouble, we are coming to realize that many others may be able to survive even in the face of abundant cowbirds and nest predators. Louisiana watert hrushes, for example, build nests in stream banks; these are difficult for both cowbirds and predators to enter. Eastern wood pewees aggressively chase intruders away from nests (Figure 3). Worm-eating warblers hide their ground nests so well (some birds even pull leaves over the nest when they leave it unattended) that cowbirds and predators rarely find them. Still other birds have a long nesting season that enables them to renest many times

Figure 5: Red-bellied woodpecker *(Malanerpes carolinus)*

and to continue nesting in late July and into August, when the cowbird breeding season is over. This is true for indigo buntings and white-eyed vireos (Figure 4), as well as for such year-round residents as northern cardinals, which do not have to fatten up for a long fall migration flight to the tropics. Although few forest birds reject cowbird eggs outright (as do American robins, northern orioles, and gray catbirds), some, such as indigo buntings and Acadian fly-catchers, on occasion abandon parasitized nests and start over.

The goal of all this research is to find out how best to reverse declines of such species as the wood thrush and to maintain the abundance of forest birds in general. Just as birds need forests, forests need birds. Bob Marquis, of the University of Missouri–St. Louis, and Chris Whelan, of Chicago's Morton Arboretum, proved the point by using large crop nets to keep songbirds, most of which are insectivorous, from a stand of white oaks. With no hungry songbirds to control them, leaf-eating caterpillars proceeded to consume so much of the

foliage that the trees grew significantly more slowly than did white oaks that were hosting a normal number of birds. The first practical step in maintaining bird populations is to preserve source areas, possibly a regional network of large forested areas, such as the Missouri Ozarks. All of the substantial songbird population sources we have identified have a high percentage of national forest at the core; these populations could disappear if these lands are turned over to any agency, state or private, that fails to maintain their expanse and integrity. Those national forests with many farms and residences—for example, the Shawnee National Forest in southeastern Illinois—may be population sinks.

The second major step in our conservation plan is to identify and restore tracts that are large enough to act as potential source habitats. One such restoration already under way is the Cache River Reserve in southern Illinois, where our field crews have been working for three years. The Cache River forests are fragmented, and cowbird parasitism and nest predation levels are high. Still, the area contains magnificent cypress swamps and miles of floodplain forests dominated by 100-foot-tall trees that grow along the natural levees of the river. The U.S. Fish and Wildlife Service, the state of Illinois, and the Nature Conservancy have developed a long-term plan to reforest an area that will eventually exceed 60,000 acres. This will provide a core area of forest buffered from farms and fields. Most challenging is the possibility of restoring the natural flood cycle of the Cache River. Research by University of Illinois graduate students Jeff Hoover and Leonardo Chapa has shown that such flood cycles are vital to many birds' floodplain habitats. Preserving and restoring forests could keep even parts of southern Illinois safe for nesting.

Excerpted from *Natural History*, July 1996.

Lake Victoria Melanie L. J. Stiassny

Freshwater ecosystems around the world exist in a complex balance, and most face severe threats. Prominent among these threats are overharvesting, pollutants (usually washed in from the land), and the introduction of foreign or exotic species. Today, the combined effect of these three types of threats has put one of the world's great ecosystems—Lake Victoria—close to death.

Lake Victoria—called the freshwater heart of Africa—is the world's largest tropical lake; it covers an area about the size of Scotland. It was once home to an astonishing diversity—more than 350 species—of cichlid (pronounced "sick-lid") species found nowhere else. Although cichlids are small fish, they were a major food resource for millions of people in the three countries surrounding the lake: Kenya, Uganda, and Tanzania.

Popular for home aquariums because of their typically vivid colors, cichlids are almost unique among fish in the way they protect their newborns from predators. They carry their newly hatched young in their mouths until the fry are large enough to have a good chance of defending themselves. The drawback of this behavioral adaptation is that it severely limits the number of offspring; in contrast, typical fish lay thousands of eggs and let them fend for themselves. This limit to the cichlid's reproduction rate makes them very vulnerable to overfishing; however, for thousands of years they thrived in Lake Victoria while playing a key role in keeping the lake's ecosystem in balance.

Around 1900, the colonial British government established a large fishing industry on the lake. They introduced gill nets, which allowed larger numbers of cichlids to be caught. For decades the lake was heavily overfished. Simultaneously, more and more local people began to settle by the lake. As land was cleared for agriculture, soil and fertilizer began washing into the lake in the runoff from the region's frequent rains. The fertilizers caused an increase in the population of surface algae. When these died, they sank, and their decomposition absorbed oxygen, reducing the oxygen available for fish living in deeper layers of the lake.

After undergoing these stresses, the lake's ecosystem suffered its severest blow of all in 1954—with the introduction of the Nile perch, a huge, voracious predator. Weighing up to 136 kilograms, the perch were introduced to give the local fishers a bigger fish to catch. Unfortunately, bigger was definitely not better. The local people had preserved the much smaller cichlids by drying them in the sun. But the only way they could preserve the much bigger Nile perch was to fry them or smoke them. Ever more forest land was cleared for the needed firewood, and this additional deforestation increased the harmful runoff into the lake while devastating the surrounding forest ecosystems.

For reasons not fully understood, the perch population remained small until the late 1970s; then it exploded—largely at the expense of the cichlids, its prey. In 1978, cichlids were about eighty percent of the biomass—the total mass of living organisms—in Lake Victoria; the Nile perch then were only about two percent of the biomass. Fewer than ten years later, the Nile perch were more than eighty percent of the lake's biomass, while cichlids were a tiny portion of the remaining twenty percent. Today less than one percent of the mass of fish caught in Lake Victoria comes from cichlids. More than half of the cichlid species may now be extinct or close to extinction.

The devastation of cichlids has led to drastic changes in Lake Victoria; even the perch's future is in doubt. The disappearance of most algae-eating cichlids allowed algae to flourish unchecked; this further depleted the lakes oxygen levels through the process described above. Today, the Nile perch and most other fish can survive only in a thin region near enough to the surface to receive sufficient oxygen from the atmosphere. The perch has virtually eaten its way through all its potential prey species except one: a species of small shrimp that appears to thrive in oxygen-depleted water. How long will the shrimp hold out? No one knows. If the shrimp goes, the whole ecosystem will collapse; Lake Victoria will die. This would be a catastrophe of enormous proportions. It could spell famine, or even starvation, for millions of people across East Africa who depend on the lake for their survival.

Although there is no way to exterminate the Nile perch, researchers are looking for ways to stabilize and possibly save the surviving cichlid populations. In 1993, conservationists introduced a Species Survival Plan for the lake's cichlids. This plan called for a widespread educational effort for people who use the lake, cichlid captive breeding programs in the United States and Europe, further assessments of the status of the remaining wild cichlid populations, and attempts to reintroduce cichlids to the lake in the future.

Will these attempts be enough to save Lake Victoria? It is too soon to know for sure, but scientists will keep trying.

Case Study

118 119

Liverwort (*Marchantia polymorpha*)

Hormonal Sabotage

Theo Colborn, Dianne Dumanoski, and John Peterson Myers

Omens

- Studying Florida's bald eagles in the early 1940s, Charles Broley followed 125 active nests and banded 150 young eaglets each year. However, in 1947 the number of young eagles suddenly began dropping sharply. Subsequently, two-thirds of the adult birds at nesting sites appeared indifferent to nesting, courtship, and mating. By the mid-1950s, Broley had become convinced that eighty percent of Florida's bald eagles were sterile.

Theo Colborn is a Senior Program Scientist at the World Wildlife Fund. Dianne Dumanoski is a writer and veteran environmental journalist who helped pioneer reporting on a new generation of global environmental issues while working at *The Boston Globe*. John Peterson Myers is a Director of the W. Alton Jones Foundation, a private environmental foundation.

- The traditional English sport of otter hunting continued unchanged into the mid-twentieth century. However, by the late 1950s, English hunters began to have trouble finding otters to hunt. Conservationists at first attributed the decline in the otter population to the pesticide dieldrin, but later work pointed to another synthetic chemical.

- In the mid-1960s, the mink industry centered on Lake Michigan—the source of fish for the ranch mink—began to falter because of the animals' mystifying reproductive problems: Females were not producing pups. Researchers eventually linked this reproductive failure to PCBs, a family of synthetic chemicals used to insulate electrical equipment.

- In 1968, naturalist Ralph Schreiber spotted gull nests with more eggs than usual on California's San Nicolas Island. Because gulls rarely incubate more than three eggs at once, Schreiber suspected that more than one female was laying in these nests. Four years later, naturalists George and Molly Hunt noticed the same phenomenon on California's Santa Barbara Island. Observing thinning eggshells in the gull colony, they suspected that these birds were suffering from exposure to the insecticide DDT. During the next two decades, naturalists observed nesting female pairs among the herring gulls in the Great Lakes, glaucous gulls in Puget Sound, and roseate terns off the coast of Massachusetts. Were the females sharing nests because of a shortage of males?

- In the 1980s, surveys showed that in some Florida lakes ninety percent of alligator eggs hatched, but at Lake Apopka the hatching rate barely reached eighteen percent—and half the hatchlings died within ten days. Reptile biologist Louis Guillette felt that the problems probably resulted from a 1980 chemical spill, after which more than ninety percent of the alligators disappeared. However, why, after the waters were again clear, were researchers still finding hatching problems, and why did at least sixty percent of the males have abnormally tiny penises?

- Over the years, Niels Skakkebaek, a reproductive researcher at the University of Copenhagen, had seen both increasing frequencies of human sperm abnormalities and decreasing typical sperm counts—while Denmark's rate of testicular cancer had tripled. He also noticed low sperm counts and unusual cells in the testes of the testicular cancer victims. Were the findings connected? Literature reviews of sixty-one studies worldwide showed that the average human male sperm count had dropped by almost fifty percent between 1938 and 1990.

The DES Case: Hormone Imitation

In April 1971, a team of doctors from Massachusetts General Hospital reported finding clear-cell cancer of the vagina in eight patients, aged fifteen through twenty-two. (This form of cancer is extremely rare and almost never strikes women under fifty.) Of the eight patients, seven were daughters of women who had taken the synthetic estrogen diethylstilbestrol (DES) during the first three months of pregnancy. In subsequent animal studies, John McLachlan and his colleagues at the National Institute of Environmental Health Sciences showed that female mice exposed to DES in their mother's womb eventually developed vaginal cancer; similarly exposed males were born with a variety of problems with their reproductive system. DES had somehow interfered with hormonal messages during the rodents' prenatal development. Subsequently, ample evidence linked DES in humans to vaginal cancer, reproductive-tract deformities, and problem pregnancies.

How do hormone imitators such as DES disrupt the endocrine system? To act, a hormone must arrive at—and bind with—the appropriate receptors in and on cells. Once joined to the receptor, hormones move into the cell's nucleus, targeting genes that "turn on" the biological activity associated with the hormone.

Figure 1: The DES Paradigm, Receptor Effects of Synthetic Chemicals

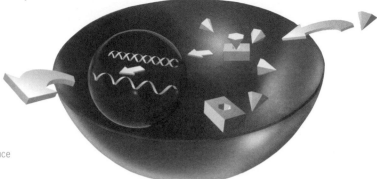

Normal response. A natural estrogen (cone) binds to a receptor within the cell (rectangle) and activates the DNA in the nucleus to produce the appropriate biological response.

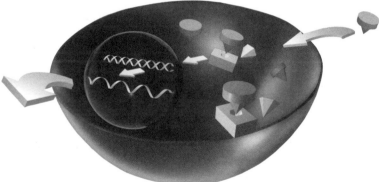

Hormone mimics. Organochlorines and other synthetic compounds can also bind to cellular receptors and elicit hormonal responses that may be stronger or weaker than those evoked naturally.

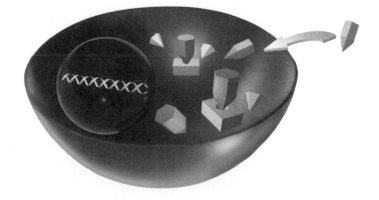

Hormone blocker. Some compounds bind with cellular receptors without detectable effects. By outcompeting the natural hormone for receptor sites, however, they suppress natural responses.

DES binds to estrogen receptors, which are found inside cells in many parts of the body, including the uterus, breasts, brain, and liver. The synthetic hormone has two troublesome traits. First, it triggers certain parts of the reproductive system more effectively than does estradiol, one of the body's own estrogens. Perhaps even more important, it manages to evade a mechanism that protects the fetus from the developmentally disruptive effects of excessive estrogen exposure. Normally, special maternal and fetal blood proteins soak up almost all excess circulating estrogen. However, they do not "recognize" DES. Consequently, DES in the fetal blood supply remains biologically active. (See Figure 1.)

Evidence suggests that other hormone imitators can also evade the body's defense mechanisms. If so, the unborn—whether humans or other animals—are vulnerable to disruption from many quarters. By 1996, researchers had identified at least fifty-one chemicals—many of them common—that disrupt hormonal control in one way or another. One hundred thousand synthetic chemicals are now on the market worldwide. About 1,000 new substances are introduced each year, most without adequate testing and review. Yearly, 5 billion pounds of pesticides are applied worldwide, not only to agricultural fields but also in parks, schools, restaurants, supermarkets, homes, and gardens. Although synthetic chemicals seem an inextricable part of modern life, their use is quite recent. Now scientists are finding evidence that different hormone-disrupting chemicals may act together to produce combined effects. Moreover, seemingly insignificant quantities of individual chemicals acting over time can have a major cumulative effect.

Here, There, Everywhere

The threat of hormone-disrupting chemicals has become known largely through a series of accidental discoveries, and none was more bizarre than one that occurred just after Christmas 1987 at Tufts Medical School in Boston. For more than two decades, physicians Ana Soto and Carlos Sonnenschein had investigated why cells multiply. Now they were experimenting with a strain of human breast cancer cells that multiply in the presence of estrogen. They were shocked to discover that these cells were rapidly multiplying in blood serum from which they had removed all estrogen. There had to be some sort of estrogen contamination, because non-estrogen-sensitive cells in other experiments were behaving as expected.

For four long, frustrating months, they checked and rechecked procedures and equipment until they discovered the source of the contamination: the plastic Corning centrifuge tubes that, until then, had never caused any problems. In July 1988, they learned from Corning that it had recently modified the composition of the plastic resin in the tubes to make them less brittle. Corning would not reveal the nature of the "new" chemical it had added to its plastic, on the grounds that it was a "trade secret." Consequently, it took Soto and Sonnenschein another year and a half of sleuthing until they had identified the "estrogen mimic" in the plastic. This substance was p-nonylphenol, part of a family of synthetic chemicals known as alkylphenols; it is commonly added to polystyrene and polyvinyl chlorides (PCVs) to make them more stable and less breakable. To learn whether p-nonylphenol acted like an estrogen in living animals—not just in a lab dish—they injected it into female rats whose ovaries had been surgically removed to prevent estrogen production. They found that the injected p-nonylphenol caused the inner lining of each rat's uterus to thicken (as would have happened if the animals had been given natural estrogen).

Searching the scientific literature, they found scattered information that heightened their concern. One study reported that the food-processing and food-packaging industries used PCVs containing alkylphenols. Another investigation found nonylphenol contamination in water that had passed through PCV tubing. The literature also revealed that alkylphenol polyethoxylates (chemicals found in many detergents, pesticides, and personal-care products) can break down into nonylphenol and other chemicals that mimic estrogens; this can happen when they encounter bacteria in animals' bodies, in the environment, or in sewage treatment plants.

Soto and Sonnenschein's findings, published in 1991, greatly surprised and alarmed even veteran investigators of hormone-disrupting chemicals. For years, scientists had thought that the health risks posed by synthetic chemicals involved human exposure to chemical residues—primarily of pesticides—in food and water. Now Soto and Sonnenschein had discovered hormone disrupters where least expected—in widespread products made from materials considered harmless and inert. Here was glaring evidence of our vast ignorance about the dangers in our everyday environment.

While Soto and Sonnenschein were chasing contamination in their lab, researchers at Stanford University School of Medicine in Palo Alto, California, accidentally discovered a different estrogen imitator—bisphenol-A—leaching from polycarbonate, an entirely different kind of plastic found in many consumer products, such as the giant jugs used to bottle drinking water. They had used polycarbonate flasks to sterilize water; in the laboratory, two to five parts per billion of bisphenol-A was enough to cause a response identical to that caused by natural estrogen.

Spurred by the Tufts researchers' report of biologically active plastics, scientists at Spain's University of Granada investigated the plastic coatings that manufacturers use to line metal cans; such linings are reportedly found in eighty-five percent of food cans in the United States and about forty percent of those sold in Spain. The brother-and-sister team of María-Fátima Olea, a food toxicologist, and Nicolás Olea, a physician specializing in endocrine cancers, ana-lyzed twenty brands of canned foods purchased in the two countries. They detected bisphenol-A contamination in about half the canned foods they analyzed, including corn, artichokes, and peas. In some cans, the concentrations of bisphenol-A were stunningly high, as much as eighty parts per billion—twenty-seven times the amount that the Stanford team reported was enough to cause breast cancer cells to multiply.

Altered Destinies

From our rapidly growing knowledge of hormone receptors—first identified in the mid-1960s—we are now beginning to understand why synthetic chemicals have such dramatic effects across an astonishing range of species. As scientists have explored hormone chemistry in various animals, they have marveled that this chemistry has remained essentially unchanged during millions of years of evolution. Whether in a turtle, a mouse, or a human, the hormonal system produces a chemically identical form of estrogen—estradiol—that binds to an estrogen receptor. The discovery of similar estrogen receptors in vertebrates as different as turtles and humans argues for a hormonal system that arose early in the evolution of vertebrates. (Turtles inhabited Earth 200 million years ago; the human genus first appeared less than 5 million years ago.)

Although research has shown that impostor chemicals bind with the estrogen receptor, it has not yet illuminated why the receptor readily accepts them. The similar effects of DDT and DES led scientists to suspect that a structural feature shared by these two molecular substances was what caused them to bind with the receptor. However, to their bewilderment, they found that the receptor also binds to chemicals with strikingly different structures.

The problem of hormone imitators is large, and it extends beyond environmental estrogens. Other classes of synthetic chemicals affect bodily processes controlled by different parts of the hormonal system, such as processes controlled by testosterone and by hormones produced by the thyroid gland. Still other synthetic chemicals inhibit the body's ability to produce such hormones as estrogen and testosterone in the first place.

The pressing question is whether humans are already suffering damage from a half century's exposure to disrupters of hormone-controlled bodily processes. Have these chemicals already altered individual destinies by scrambling the chemical messages that guide bodily development? Many of those familiar with the scientific case believe that the answer is yes. Nevertheless, whether hormone-disrupting compounds are also having a broad impact across the human population is difficult to assess and even harder to prove.

The chemical age has created products, institutions, and cultural attitudes that require synthetic chemicals to sustain them. The task that confronts us over the next half century is one of redesign. We must find safer ways to meet human needs. As we work to create a future where children can be born free of chemical contamination, our scientific knowledge and technological expertise will be crucial. Nothing, however, will be more important to human well-being and survival than the wisdom to appreciate that however great our knowledge, our ignorance is even greater.

124 125

Gonium *(Gonium perctorale)*

Massive *Porites* colony, Agincourt Reef,
Outer Great Barrier Reef, Australia.

Reefs in Crisis Barbie Bischoff

Humans have harmlessly harvested the rich wildlife on coral reefs for thousands, perhaps hundreds of thousands, of years. But in the late twentieth century, human pressures—including a population explosion and migration to coastal areas—have placed reefs at risk. In the Caribbean Basin, the population has quadrupled since 1960, and seventy-five percent of the people live within six miles of the coastline. Natural events, such as El Niño, have also played a role in the decline of reefs. And tourism has been both blessing and bane.

According to a report issued by the International Coral Reef Initiative, tourism accounts for more than fifty percent of the gross national product of several Caribbean countries; this provides an economic incentive for reef protection. But more visitors means more coral collecting and more damage caused by swimmers, divers, and boat anchors. Moreover, the clearing of land to make way for hotels and homes has greatly increased the rate of shoreline erosion. Without the natural filter provided by wetland vegetation, soil pours into the sea, blocking the sunlight vital to corals and choking the pores of sponges. Because of poor agricultural practices and wetland destruction to make way for an increasing population, reefs are getting large doses of fertilizers from agricultural runoff. The nitrogen and phosphorus in the compounds have overnutrified the water, a condition called eutrophication, and allowed fibrous and fleshy lettuce-like algae to take hold along the reefs.

Irresponsible tourism is a significant problem for reefs, especially in the Caribbean. Here, well-intentioned tourists stand on the live coral, which damages the tissues of the undersea colonies, often beyond repair and sometimes results in the destruction of entire reef tracts.

The creatures that are supposed to eat algae cannot keep pace with the accelerated growth and often abandon the reef in search of a more balanced environment. Eventually, the eutrophic reef becomes a ruin as the algae thrive, starving the coral of the sunlight it needs.

Coastal waters are degraded off southern Florida, Haiti, Cuba, the Dominican Republic, and Veracruz, Mexico. Haiti's case is acute, because only one percent of native coastal vegetation remains, and sewage treatment plants have yet to be built. Contamination from fossil fuels, industrial chemicals, and pesticides—as well as domestic and animal waste—is also a problem throughout the Caribbean.

Like other developing nations, some Caribbean countries are forced to survive by overexploiting their own resources for the global market. And coral reefs, which occupy only about 0.2 percent of the world's oceans, supply about 9 million of the 80 million tons of fish harvested worldwide each year. Some harvesting methods, such as mechanical dredging or large-scale poisoning, irreparably damage the reefs. And overfishing has made the queen conch (a large, spiral-shelled mollusk), the spiny lobster, the whelk (a marine snail), the red snapper (a fish), and the Nassau grouper (a fish) commercially extinct in many localities. The once-abundant jewfish, a grouper, has virtually vanished from the Caribbean. In Haiti, populations of larger reef fish and lobsters are "crashing" (decreasing to very low levels) because many of these animals are taken from the sea before reaching reproductive maturity.

Commercially desirable fish and crustaceans aren't the only casualties. Illegal sale of turtles is common in the Dominican Republic, the Bahamas, and Mexico. Tourists' fancy for such souvenirs as shells, coral skeletons, and other curios has depleted black coral and mollusks. Meanwhile, a growing aquarium trade has overharvested smaller, ornamental fishes.

The incidence of coral disease is also climbing. And in 1997, El Niño was unexpectedly intense. This caused ocean water temperatures to rise in many reef locations. This in turn produced the worst bleaching (expulsion of the colorful algae that live within corals) seen in the last decade. In 1998, similar El Niño–caused bleaching occurred in the Great Barrier Reef off Australia's northeastern coast.

There is the potential for good news, however. Sanctuaries are beginning to change. Formerly designed as simply small, totally protected areas, they had little impact on the health of reefs. Now the trend is to divide large areas into zones for distinct uses, such as fishing, tourism, shipping, defense, collecting, scientific research, and hunting and fishing by local residents. Florida Keys National Marine Sanctuary is among those successfully adopting this approach. Many parks throughout the Caribbean, however, don't have the funding necessary for maintenance and enforcement.

While scientists have long recognized the importance of the land-water connection to reef health, implementation of good land management techniques is only beginning in many areas. Development within parks such as the Sian Ka'an Biosphere reserve, in the Yucatán Peninsula, is regulated. In the United States, Florida now requires barriers to control sediment generated by construction projects.

In 1997—the International Year of the Reef—a public awareness campaign, conducted on a grass-roots level, attempted to inspire local residents to protect their nearby reef ecosystems. Within the last few years, various organizations have begun to sponsor monitoring programs, mapping expeditions, scientific research, and focused conservation and management efforts.

Excerpted from *Natural History*, December 1997–January 1998.

Giant sequoia leaves (*Sequoiadendron giganteum*)

Section Three: **Saving Biodiversity: Strategies and Solutions**

Branching and plate corals, Wilson Island,
Great Barrier Reef, Australia.

Introduction Michael J. Novacek

Efforts to mitigate the current loss of species and habitats should be multidisciplinary in scope and global in scale. This section highlights efforts to save biodiversity, opening with a prescient essay, published in 1948, by Fairfield Osborn. In "The New Geologic Force: Man," he called on us to transform our subjugation of natural systems of Earth into effective stewardship. In more specific ways, Curt Meine outlines the steps in conservation management that have shifted from management of individual species to the preservation of habitats and the ecosystems they represent. Such well-informed management requires scientific knowledge equitably distributed among countries. Joel Cracraft describes the challenge and urgent need for training young scientists and policy makers in countries that are technologically behind but at the same time claim most of the world's biodiversity. Our effectiveness in mustering a truly international effort will be critical to any measure of success in stemming biodiversity loss. For urgent scientific response and training, space-age technology is useful and welcome. For example, Prashant Hedao describes the application of satellite-based Global Information Systems (GIS) in detecting environmental contours and disruption over a broad regional scale with surprising detail. Good science also underpins well-applied conservation. Thus, some practices involve not only preservation but also forest restoration as described by Dan Janzen in a project leading to the successful reestablishment of tropical dry forest.

In our global economy, solutions to the biodiversity crisis have been motivated by more than a recognition and respect for nature. Graciela Chichilnisky explains that natural resources have an economic value that often exceeds their market price. She argues that global prices paid for such resources do not reflect the irreplaceable loss that their extraction represents to developing countries, and she suggests the need for the economy to develop monetary values for environmental resources. At more local levels, communities can take steps to conserve biodiversity by using natural resources in a sustainable fashion. Nick Salafsky notes that such a model is particularly important in developing countries, where national parks and reserves common to many countries are overly restrictive to local indigenous populations. This gives rise to an essential question. What do we mean when we describe rain forest products and other natural resources as "sustainable"? Charles Peters addresses this issue in his essay, aptly entitled "No Free Lunch in the Rain Forest." As he points out, sustainable products are derived at a rate and intensity that allow the unhampered reproduction and replenishment of a species. The sustainable approach maintains an equilibrium for the ecosystem that benefits all species, including humans who utilize these products. Such practices are essential; for example, rain forests broadly lumbered or burned rarely recover. The delicate balance required for a sustainable program is a challenge as we enter this uncertain century. But the case studies described in this section, as well as the profiles of people dedicated to containing the biodiversity crisis, are meant to help spark the broader and more intensive international effort that is so urgently required.

To explore feasible solutions to mitigating biodiversity loss I pose the following questions:

What are the consequences of modern technology?

Fairfield Osborn, President of the New York Wildlife Conservation Society from 1940 to

1968, argues, in an extract from a book published in 1948, that our abilities to live in almost every conceivable habitat, to increase our food production, and to communicate with others around the globe, bring with them a responsibility to act as stewards of the Earth.

What is the history of resource management in the U.S.?

Curt Meine, Director of Conservation Programs at the Wisconsin Academy of Sciences, Arts and Letters, outlines the historical changes in national conservation biology and wildlife management.

Why is there a need for systematic biologists?

Joel Cracraft, Curator-in-Charge of the Department of Ornithology at the American Museum of Natural History, Co-Curator of the Museum's Hall of Biodiversity, and on the Scientific Steering Committee of the Systematics Agenda 2000 International, asserts that international scientists, in collaboration, must identify, name, and describe as many of the world's millions of species as possible to enable further insight into the potential benefits of biodiversity.

How can a tropical dry forest be restored?

Daniel H. Janzen, Professor of Biology at the University of Pennsylvania, describes how he and his colleague, Winnie Hallwachs, proposed and initiated a project for setting aside a large area of forest remnants, how this project was managed, and how it resulted in the successful reestablishment of tropical dry forest.

Does the free market do a good job of setting prices for natural resources?

Graciela Chichilnisky, UNESCO Chair of Mathematics and Economics and Professor of Statistics at Columbia University, believes that the global prices paid for natural resources do not reflect the loss that their extraction represents to developing countries and suggests that economists should develop monetary values for environmental resources.

How can new forms of mapping help preserve biodiversity?

Prashant M. Hedao, of the Environmental Systems Research Institute, Inc., in Redlands, California, summarizes the historical uses of mapping, and describes how conservationists can use this technology to advocate for measures to prevent further loss of biodiversity.

How do we reconcile the loss of biodiversity with the needs of an ever-expanding human population?

Nick Salafsky, one of the founders of Foundations of Success, proposes that biodiversity conservation and sustainable development go hand-in-hand and demonstrates this with the use of several models.

What do we mean when we describe rain forest products as "sustainable"?

Charles M. Peters, the Kate E. Tode Curator of Botany at the Institute of Economic Botany at the New York Botanical Garden, advocates a close look at such designations to ensure that rain forest products are harvested in a manner that allows for their continued survival.

Fairfield Osborn

The New Geologic Force: Man

Fairfield Osborn

Before taking a closer look at this Earth-home in which more than two billion human beings are trying to work out their survival, let us first consider the drastic change in the world picture resulting from the spectacular and cumulative series of modern inventions. In more ways than one the Earth is far from being what it was—even the other day.

Fairfield Osborn, naturalist and conservationist, was President of the Wildlife Conservation Society (1940–68) and President of the Conservation Foundation, a group designed to initiate and advance research and education in the entire field of conservation. He wrote several influential books presaging the crisis in biodiversity on global environmental issues, including *Our Plundered Planet*, 1948. His father, Henry Fairfield Osborn, was President of the American Museum of Natural History for many years.

Space, as well as time, is relative and our conceptions regarding both are constantly changing. Modern thought has come to recognize that space and time are closely interrelated. This realization has produced, within the present century, a new theory concerning the cosmic scheme, including even a new definition of infinity. While a concept having to do with the remote boundaries of the universe is not directly pertinent to the consideration of man's relationship with nature and, more particularly, with his own living spaces, we find that within recent decades our entire point of view concerning this Earth-home of ours has also undergone a major change. The remarkable development of the technical sciences has caused the Earth, among other things, to become constantly smaller....
The Earth was referred to as one of the minor planets belonging to a star of moderate size. In itself this description is of no particular import, principally because it involves a consideration of size, or space, without relationship to time. Speed of communication and rapidity of transportation are eliminating distance. Yesterday, we began using the moon's surface as a reflecting board for radar messages to the other side of the Earth. Not so long ago—in the sixteenth century to be exact—a round-the-world trip, or message, took more than three years. Now, it is possible to dispatch a message around the Earth in a few seconds and to travel around it in less than four days.

So it is that the Earth is constantly becoming smaller, or rather our knowledge of it is leading us to think of it as diminishing rapidly, which, after all, amounts to one and the same thing. As a consequence we are now thinking of mankind in terms of a world society. The boundaries or barriers between localities, nations, even continental populations, are dissolving. From a social or political point of view, the process is slow and intensely painful—

marked by rancor and bitterness, jealousy and warfare of unbelievable destructiveness. From a physical point of view, that is insofar as human beings themselves are concerned, there is no actual change. It is merely that people the world over are coming to realize the essential unity of mankind.... The likeness of all human beings, the fact that from a biological point of view no nations or races are disparate, that human beings throughout the world are of one species—or at most divided into groups or subspecies, all closely similar—makes this change in our conceptions of human civilization unavoidable and inevitable. Further, due to the existence today of worldwide systems of commerce, combined with new and so-called higher standards of living, all nations are dependent on others in varying degrees for products, materials or goods that have become a necessary part of everyday living for most of the people on the face of the Earth.

The conditions—whether material or social or even ideological—which exist among peoples in one section of the Earth now have a bearing on the lives of peoples of far distant nations. No longer is an American unaffected by the trends of living conditions of other peoples, whether those of a country in the Western Hemisphere or even those on the exact opposite side of the Earth. No longer can three million people in India die of starvation, as they did in 1943, without a specific and cumulative effect on an Englishman in Sussex. The spoiling of the land and the ensuing destruction by floods in the great Yellow River Valley of China sooner or later, in one manner or another, impinge on the well-being of peoples 1,000 horizons away. The peoples of the Earth, whether they will it so or not, are bound together today by common interests and needs, the most basic of which are, of course, food supply and other primary living requirements. These come, all of them, from nature and from nature alone—from the forests, the soils, and the waterways. Man's problem in his earliest, dimmest, most faraway days was obtaining a living from these elements. The wheel of human destiny seems to turn, but the basic facts of life remain constant. Man's initial problem is still with him— can he obtain a living from nature? The population of the Earth has increased almost five times within the last three centuries and doubled even within the last century. Human civilization has permeated virtually every living area of the Earth's surface…. Vast fertile areas in various parts of the Earth have been injured by man, many of them so ruined that they have become deserts and uninhabitable. In such places, flourishing civilizations have disappeared, their cities buried under wastes of sand, their inhabitants scattering to new lands. But now, with isolated and inconsequential exceptions, there are no fresh lands, anywhere. Never before in man's history has this been the case.

Actually the Earth, from the point of view of its use as a place to live in, is far smaller than our minds picture it. We retain, even after we are grown up, our first childhood impressions of the vastness of the "great round Earth." We are apt to forget that almost three quarters of the Earth is covered by water and that at least one half of all the land is uninhabitable because it lies in the polar regions or is extremely mountainous or is desert land. Consequently, there remain only about twenty-five million square miles, equivalent in round figures to about sixteen billion acres, that can be thought of as originally favorable to habitation by man. Divided by the number of people alive today, this would mean a theoretical maximum of less than eight acres of naturally habitable land for each human being if the total habitable regions were divided equally. We must of course envisage all types of land, including those covered by forest, lying in grasslands, or those favorable to the cultivation

of crops. In the present state of statistical knowledge we cannot estimate with exactness the proportions of the entire habitable areas of the Earth that are devoted to the different uses to which man has put them. We do know that a very large proportion of the originally habitable areas have already been so misused by man that they have lost their productive capacity. Extensive areas of man-made deserts— sterile, barren, beyond reclamation—exist on every continent. Innumerable other areas, all over the Earth, have been robbed of so much of their value that they are barely worth cultivation; the products from these lands possess little energy content; the people are undernourished.

As to the remaining amount of land that can be used for cultivation, the productive soil of the world is now so limited that it is estimated there are not more than four billion acres of arable land left to fill the needs of more than two billion people. A study made by the United States Department of State reports that the area of cultivated land in the world before the outbreak of World War II was somewhat less than two and one-half billion acres. If one takes the larger figure of four billion acres, representing the area of land estimated as now available or suitable for cultivation, it means that there are less than two acres per capita. Contrasted with this, it is a generally accepted computation that two and one-half acres of land of average productivity are required to provide even a minimum adequate diet for each person. Many countries have less than an acre of productive land per capita. No wonder there are worldwide shortages, and that the people of a number of nations are facing starvation.

Blind to the need of co-operating with nature, man is destroying the sources of his life. Another century like the last and civilization will be facing its final crisis. While there is a growing, frightened awareness of the oncoming peril—while in some countries corrective steps of almost sufficient vigor are being taken—the fact remains that the turning point of recovery and reclamation has not yet been reached either in this county or elsewhere. The issue has not yet been met. The third of the Four Freedoms, "Freedom from Want," Dumbarton Oaks, the San Francisco Conference, the United Nations meetings—all of these reachings of the human mind and spirit for a better world will prove meaningless and futile until this issue is met; until, through worldwide planning, we first protect what remains and then take steps, wherever possible, to start back on the long, slow road of reclamation.

This road can be found and traveled only if there is general understanding of the problem that confronts us and overall programs, international as well as national, are devised to cope with it. Before considering some of the interrelated processes by which nature provides the essentials that are needed for man's survival, it would be well to give thought to the amazing increase in human numbers, especially within recent centuries.

136 137

Liverwort *(Bryopteris filicina)*

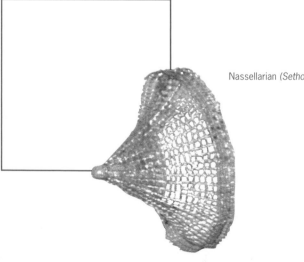

Nassellarian *(Sethophormis eupilium)*

Jane Goodall in Gombe National Park in Tanzania, East Africa.

Jane Goodall: **Profile**

Dr. Jane Goodall's path-breaking African field studies of chimpanzee social behavior forever changed the way we view our closest evolutionary relatives. By causing her readers to empathize with the wild chimpanzees she got to know on an almost personal basis, Dr. Goodall provided a compelling moral argument for allowing these intelligent primates to flourish in their natural habitats.

Among her many other conservation activities, Dr. Goodall has created an organization called Roots & Shoots. Roots & Shoots is a symbolic name. "Roots creep underground everywhere and make a firm foundation," Dr. Goodall says. "Shoots seem very weak, but to reach the light, they can break open brick walls. Imagine that the brick walls are all the problems we have inflicted on our planet. Hundreds of thousands of roots and shoots, hundreds of thousands of young people around the world, can break through these walls. You can change the world." Today, as members of Roots & Shoots, thousands of children in more than forty-eight countries are planting trees, protecting species, and otherwise working to improve the environment.

Sea anemone *(Cribrinopsis fernaldi)*

Humpback Whale Conservation Genetics Project, Madagascar

Howard Rosenbaum

Humpback whale breaching off the Madagascar coast.

Well known for its efforts to protect such terrestrial mammals as lemurs and other animals found nowhere else, Madagascar is now focusing attention on conserving its native marine mammals. A significant population of humpback whales *(Megaptera novaengliae)* inhabits or migrates through Madagascar waters. However, little information exists on their distribution, abundance, or movements. In fact, the status of humpbacks throughout most of the western Indian Ocean is poorly known, but now improving.

This conservation genetics project, sponsored by the American Museum of Natural History's Center for Biodiversity and Conservation and the Wildlife Conservation Society, is studying the humpback population that frequents Antongil Bay, which is off northern Madagascar. Our data show this area to be a major breeding and calving ground.

A major focus of the study is to determine the population size of these whales and to learn about the relationship between whales in the groups making up this population. This is done by taking photographs of distinguishing markings that are unique to individual whales. Examination of the characteristic scars and coloring on their tail flukes allows for the identification of individuals. So far, we have identified about 600 whales, including fifty newborn calves.

Another research goal is to understand the genetic relationships of the humpback population. By obtaining skin samples, we can examine the whales for genetic similarity, both within the waters of Madagascar, throughout the Indian Ocean, and across the Southern Hemisphere. A better understanding of how humpbacks migrate and mix with other populations will help in the design of conservation measures to protect them.

A primary goal of this project is to improve marine conservation research and management within Madagascar. To this end, the Center trains Malagasy conservation biologists and students in field techniques, and provides financial support for students to return to the American Museum of Natural History for further study and analysis of project data. The development of a well-trained force of Malagasy conservationists is essential to the long-term success of any conservation program.

As with all Center research programs, this humpback project promotes the collection of field data and their effective application to a conservation problem. In this case, our efforts will help in the design of marine sanctuaries and reserves, the regulation of ecotourism activities (in this case, "whale watching"), and the integration of local perspectives and concerns regarding whale conservation."

Mexican beaded lizard *(Heloderma horridum)*

Portrait of Aldo Leopold

Conservation Biology and Wildlife Management in America: A Historical Perspective

Curt Meine

"All through college I was trained to create edge, edge, and more edge. Now all I'm hearing is that edge is bad!" The words were those of a district-level wildlife manager in the U.S. Forest Service. He and two dozen colleagues were attending an agency-sponsored continuing education program designed to keep them up to date on innovations in habitat management.

Curt Meine is Director of Conservation Programs at the Wisconsin Academy of Sciences, Arts and Letters.

However, after several days of patient participation, the agency veteran could no longer contain his confusion. What he was hearing (this was the late 1980s) and what he had been told in college (perhaps twenty years before) simply did not jibe.

His exasperation revealed much, not only about recent changes in our view of edge effects but also about longer-term changes within (and surrounding) the field of wildlife management generally. The immediate source of his confusion could be guessed. Probably none of his college instructors had explained—perhaps they did not realize themselves—the origins and development of the edge-effect concept and its application in wildlife management. (For ecologists, an "edge effect" refers to the changes in environment conditions that occur at the boundary between two habitat types. These changes may include changes in the species present.) None cautioned that the creation of edge habitat was a management tool, and as with all tools, its appropriate and effective use depended on timing, location, and ecological context. None mentioned that, in the 1930s, maximization of edge habitat was a "progressive" technique in the then-new field of game management. (Wildlife managers used this approach to restore game populations in the Midwest. After decades of intensive agricultural development, this landscape retained little edge, little cover of any kind, and only insignificant remnants of its originally extensive living communities.) None of his instructors foresaw how the too-eager use of the "edge habitat" tool, especially in forest settings, could have harmful effects on species that normally lived deep within forests.

The manager's complaint revealed a still deeper frustration. Beyond concerns about the proper application of this or that technique, he was confronting the rapidly changing role and context of wildlife management as it entered the 1990s. At a time when human impact on the world's ecosystems has increased, the reconsideration of edge effects has been only one outward sign of a more basic reappraisal of means and goals in the effort to conserve wild places and the plants and animals they contain. The emergence of conservation biology itself, with its special emphasis on protection and maintenance of biological diversity, has been another important indicator of this reappraisal.

Seen from one angle, conservation biology directly challenges many assumptions and priorities that have guided wildlife management for five decades. Wildlife management has heavily emphasized a narrow range of species, typically game animals or other animals (typically higher vertebrates) that are economically valuable, particularly attractive, or symbolic (such as the bald eagle, a symbol of the United States). It has also usually focused on single species in research and management, thus underestimating the importance of broader, system-wide approaches. Education and training programs have stressed the development of technical skills while downplaying conceptual clarity and intellectual flexibility. Finally, wildlife management has had a relatively rigid disciplinary framework that carries over from the classroom to the agency department—and ultimately to the landscape. These tendencies are not unique to wildlife management. They also occur within agriculture, forestry, range management, fisheries management, and other resource-related professions.

The roots of conservation biology lie in many fields, within and beyond the natural sciences. Wildlife management is only one of these, but its contributions have been disproportionately important. It was the arena in which biological knowledge and ecological principles were first applied systematically to the conservation of organisms and their natural habitats. As such, it played a leading role in advancing conservation beyond the point where we measured success solely in human economic terms (whether that measure was board-feet produced, deer "harvested," fingerlings released, acre-feet retained, or tourists admitted). In this way, it helped initiate the process—still far from complete—that would redefine conservation as the effort to protect, manage, and restore healthy and diverse ecosystems. Seen from this perspective, conservation biology is not a radical departure from the past but instead a further stage in conservation's continual evolutionary process. Along the way, conservation biology has given "traditional" wildlife management the opportunity to return to its roots and to revise or reaffirm many of its founding (if sometimes neglected) principles.

When wildlife protection emerged as a distinct profession in the 1930s, it represented a significant departure from conservation's status quo. As Aldo Leopold noted in his pioneering text *Game Management* (1933), "the thought was that restriction of hunting could 'string out' the remnants of the virgin supply [of game animal populations] and make them last a longer time….Our game laws…were essentially a device for dividing up a dwindling treasure." Leopold introduced the profound idea that we could best perpetuate populations of wild animals and plants through the active study, protection, and, where necessary, restoration of their habitats. He called on science "to furnish biological facts" and "to build on them a new technique by which the altruistic idea of conservation can be made a practical reality."

Summarizing several key developments can provide some sense of how this new approach transformed wildlife conservation in America in the crucial decade of the 1930s.

- In 1930, nearly all those involved in the management (as distinguished from the study or protection) of wild animal populations focused on game species. In 1936, the one-word term "wildlife" came into common usage, signaling the broadened concern of the field. By 1940, "wildlife" was standard terminology; for many people, wildlife included not only "nongame" vertebrates but also invertebrates and plants as well.

- Before 1930, "management," such as it was, entailed mainly captive breeding programs; the killing or removal of species that preyed on farm, ranch, or game animals; tighter legal restrictions on hunting; and the unsystematic creation of refuges and sanctuaries to protect specific species in specific locations. By 1940, the basic shift in approach was complete, and primary emphasis was now placed on the provision of suitable habitats.

- In 1930, only a small group of people, mainly in academia, understood the science of ecology. By 1940, ecology was the cornerstone of wildlife management.

- In 1930, there were no textbooks, journals, or professional organizations devoted exclusively to the emerging field of wildlife management. By 1940, this field had its text (Leopold's *Game Management*), its journal (*Journal of Wildlife Management*), and its professional society (the Wildlife Society).

- In 1930, one could count on the fingers of one hand the number of research projects designed specifically "to furnish biological facts" relevant to the conservation of wildlife. By 1940, a national system of financial and institutional support for wildlife research (the Cooperative Wildlife Research Unit program) had been established.

- In 1930, opportunities to study wildlife management were virtually nonexistent, confined (at best) to an occasional lecture in a forestry or agriculture class. By 1940, courses and whole departments devoted to wildlife management were in place in dozens of universities (particularly the nation's land-grant universities, which were built, at least in part, on land "granted" to them by the government).

- In 1930, few people appreciated the connections among wildlife ecology and management, other basic and applied sciences, and economics, philosophy, and other fields. By 1940, spurred on considerably by the new generation of "wildlifers," people were beginning to grasp the broad implications of conservation.

Even as wildlife management was securing these professional footholds, its conceptual foundation continued to broaden. During these years of ferment, the focus had expanded well beyond Leopold's original aim of "making land produce sustained annual crops of game for recreational use." The shift toward the more inclusive term wildlife reflected not only an interest in a wider spectrum of species but also a deeper realization of the widespread importance of the science of ecology. Leopold himself saw this clearly. In a 1939 address to a joint meeting of the Society of American Foresters and the Ecological Society of America, he described ecology as "a new fusion point for all the natural sciences." He noted that ecology challenged traditional notions of placing a value on particular species, even as it high-lighted the basic importance of biological diversity. "No species," he proposed, "can be 'rated' without the tongue in the cheek. The old categories of 'useful' and 'harmful' have validity only as conditioned by time, place, and circumstance. The only sure conclusion is that the biota as a whole is useful, and biota includes not only plants and animals, but soils and waters as well."

Leopold was hardly alone in this realization. In every field of natural resource management, there were "dissenters" (to use Leopold's term) who reached the same conclusion: Understanding of natural phenomena and human environmental impacts came not simply through the division of reality into smaller and smaller bits—the method of reductionist science—but through greater attention to the connections and relationships in nature at various scales of time and space. For the conservation professions, this had important practical implications. One could not simply manage soils, timber trees, game animals, or any other "resource" as separate entities; one also had to treat the ecological processes that kept the system as a whole healthy. This meant that, despite departmental and disciplinary labels, integration was essential to all conservation work.

This line of thinking would endure even the tumult of World War II. As an inherently integrative undertaking, wildlife management was partially immune to the postwar trend toward extreme specialization. Through the 1950s, there was close, active, and regular interaction among academic ecologists, other biologists, and the applied wildlife management programs in the universities and the state and federal governments.

By the end of the 1950s, however, even wildlife management began to suffer from "hardening of the categories." As noted in 1989 by Fred Wagner, new directions in the underlying sciences "would send academic ecology and applied wildlife management down somewhat different paths and dissolve the close association of previous decades." In the following decades, these widening gaps— between theoretical and applied scientists, between scientists and managers, between departments in the agencies and universities— would make consensus ever more difficult. Simultaneously, threats to the biota, at all geographic scales, intensified. In short, the "glue" that first allowed wildlife management to come together and stick together—an expanding appreciation of biological diversity and ecosystem processes, broad training in the natural sciences, collaborative research projects, and integrated approaches to resource management—was allowed to break down. For a generation, fragmentation would become increasingly evident, not only in the modern landscape but also in the modern mindscape.

The quickening pace of environmental degradation and biological impoverishment in the 1960s and 1970s would outstrip the ability of the various conservation-related sciences, acting in isolation, to respond. In a world beset by complex, large-scale, interrelated environmental concerns—including deforestation, air and water pollution, global climate change, human population growth, and misguided international development projects—wildlife management as generally practiced seemed less and less relevant or responsive.

The newly energized environmental movement sought to confront these trends through ambitious conferences, management programs, and legislative initiatives at the national and international levels. Yet these measures alone could not reverse the trends. Ultimately, we could attain conservation goals only through understanding and changing the entrenched patterns of resource use that threatened plant and animal populations, degraded their habitats, and disrupted the functioning of ecosystems. This was, in many ways, the proper domain of wildlife management. However, to respond, the profession has had to rethink its priorities, broaden its mission, and reintegrate itself with the other resource management professions.

For many in wildlife management, that process has in fact gone on under the name of conservation biology.

The rise of conservation biology has all but inevitably provoked defensiveness by some in the "traditional" conservation fields. However, it has also allowed many—from the agency head to the district-level wildlife manager—to step beyond, and return to, their respective areas of expertise with a deeper sense of their professional roots, their shared goals, and the special contribution they can make to the common cause. Conservation biology treats the world not as a collection of separate specialties but as an interconnected whole to which each specialty can bring emphasis, insight, and perspective. In this, it is not primarily a challenge to wildlife management. Rather it is a fulfillment of the conservation vision to which wildlife management has always given so much. As wildlife management redefines its own future role accordingly, it can take justifiable pride in its historic efforts to promote what Leopold called "that new social concept toward which conservation is groping."

Pine snake (*Pituophis melanoleucus*)

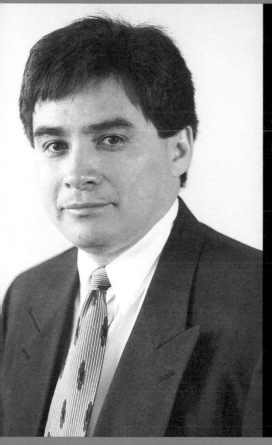

Jaime A. Pinkham: Profile

The Nez Perce tribe recently reacquired a portion of their ancestral homeland in north-eastern Oregon that they had been forced off in the 1870s. According to Jaime A. Pinkham, Manager, Department of Fisheries Resource Management for the Nez Perce Tribe, "Returning to those mountains was like a homecoming. The history and the spirit of the Nez Perce never faded from those mountains."

Before being elected to the Executive Committee, Pinkham was the manager of the tribal Department of Natural Resources. "The management of our natural resources is designed to meet the demands of today's society on a sustained yield, while at the same time providing cultural protection and preservation of the diversity of the landscape."

Educated as a forester, he acknowledges that there is still much to learn about the land. "Not only do we have to learn how to better care for the land, we also need to preserve it in a way that protects its many secrets that may someday hold the answers to some of the world's problems."

Pinkham sees the protection of biological diversity and cultural diversity going hand in hand. "The Nez Perce are land-based people. A healthy environment is fundamental to our way of life and traditions. And the survival of certain resources is sacred to the expression of our beliefs."

Managing the Biosphere:
The Essential Role
of Systematic Biology

Joel Cracraft

By ratifying the Convention on Biological Diversity, about 171 nations have signaled their intention to strive for a sustainable world—an environmentally healthy world that can continue to produce the goods and services required by its human inhabitants. Everyone agrees that this is an essential goal, but we may have greatly underestimated the difficulty of accomplishing it.

Cur
Joel Cracraft is Curator-in-Charge of the Department of Orinthology at the American Museum of Natural History and Co-Curator of the Museum's Hall of Biodiversity, as well as on the Scientific Steering Committee of the Systematics Agenda 2000 International.

Around the world, people currently use tens of thousands of species to meet their daily needs. Simultaneously, however, they are causing Earth's ecosystems to deteriorate, and Earth's biodiversity to decrease, at an accelerating rate. This, in turn, threatens the health and well-being of people everywhere. Reversing these related trends, or even slowing them significantly, presents the world with many formidable challenges. Do we even have sufficient scientific information about the biosphere to manage it sustainably? Scientists have universally declared that the answer to this question is, no, we do not.

Indeed, we lack sufficient information about the ecosystems in the scientifically sophisticated developed nations. However, most of the world's nations are poor, lacking even the beginnings of a decent scientific infrastructure. The tropical countries, which house an estimated eighty percent of the global biodiversity, have only about six percent of the world's scientists. Clearly, we need to increase our capacity to understand and manage the world's ecosystems—particularly those in species-rich countries. As the cornerstone for such a program, we need to increase capacity in the scientific area of systematics.

What is Systematics and Why is it Important?

The biodiversity sciences encompass all the disciplines concerned with understanding, conserving, and sustainably using biodiversity. The "core" of biodiversity science consists of disciplines whose research focuses on whole organisms; the most important of these disciplines are ecology, population biology and genetics, the study of behavior, and systematics. Systematics, also called "biological systematics" or "taxonomy," is perhaps the most fundamental of the bio-diversity sciences because it documents the nature and extent of Earth's biodiversity.

Like most sciences, systematics can be defined by its research questions and objectives. Systematists (sometimes called taxonomists) describe, name, and classify the various species of living things. The basic unit of biological classification is the species—one or more populations that can be distinguished from other such populations on the basis of some diagnostic features or characters. The next-higher classification group is the genus, which includes closely related species. Then, in ascending order, the classification groups above the genus are the family; the order; class; phylum (for animals) or division (for plants, fungi, and protists); and, at the top of the hierarchy, kingdom.

Hierarchial classifications are based on an understanding of evolutionary relationships. Phylogenetics is the discipline that attempts to understand these evolutionary (historical or "genealogical") relationships among species and groups of species.

Knowledge of systematics is crucially important for society and for managing our natural resources, but we often take this importance for granted. Accurate identifications of disease organisms such as bacteria or parasites, and of organisms that spread disease-producing organisms (for example, some insects and rodents), are essential for ensuring human health. The world food supply is constantly under attack from many different kinds of pests, and their identification is critical for developing strategies to combat them. Likewise, systematists exploring in the field may discover new species that produce novel substances that have useful medicinal properties. Similarly, they may find species whose genes might help to improve crop yields or increase the resistance of crop species to disease.

As with all scientists, systematists have special requirements to undertake their research. Their

most important need is access to specimen-based collections housed in systematics research institutions of various kinds. The world's collections contain 2 billion to 3 billion specimens. These collections, which are society's only permanent record of Earth's biodiversity, take many forms. Among these are natural history museums, herbaria, frozen tissue collections, seed banks, bacterial type culture collections, and, for some types of studies, living material in zoos, aquariums, and botanical gardens.

Systematics collections serve a much broader —and possibly more important—role than just providing a basis for scientific research. Through their exhibits and other programs, collection-based institutions such as museums and botanical gardens play an essential function in educating the public about the benefits of, and threats to, biodiversity. These institutions also provide public lectures, minicourses, and other formal educational programs to people ranging from young schoolchildren to people receiving

professional or paraprofessional training. Little of this public education could take place but for the scientific collections that form the foundation of this educational outreach.

Global Patterns of Systematics Capacity

Countries differ greatly in their capacity to undertake research in systematics. Among the world's major regions, Europe and North America have the greatest numbers of museums, zoos, aquariums, and botanical gardens. South America, Australasia, and Africa have fewer, and the number per region decreases in that order. Moreover, within some of these latter regions, one country—for example, South Africa within Africa, or Australia within Australasia—dominates the statistics. Many developing countries in Latin America, Africa, and tropical Asia lack even the beginnings of such a capacity; many developing countries have no institutions housing botanical or zoological collections.

The implications of these observations are profound: Countries lacking these institutions lack the capability to preserve a permanent record of their biodiversity. Because these institutions are necessary centers for training and teaching, countries lacking them are also failing to build the necessary expertise to manage their natural resources.

Leading experts on biodiversity recognize an ever deeper tragedy in these patterns: even in the wealthy countries, the systematics capacity is too small to meet the overall global need. In fact, systematics capacity in the developed nations is barely adequate—if not totally inadequate—to meet these nations' own demands for systematics information.

We have no choice, then, but to develop systematics capacity in all nations, particularly in the species-rich regions where need is greatest.

Advancing Systematics Knowledge Around the World

Yet, how can we accomplish this? First, we need to understand that improving any aspect of science capacity depends on a parallel improvement in the overall well-being of society. Many countries are struggling to meet the basic needs of their people—for food, shelter, education, and health care. Under these conditions, developing science is difficult and is often seen more as a luxury than as a long-term investment in the society's future. Thus, improvements in people's quality of life—especially in education, health care, and political justice for women—are prerequisites for advances in science, including biodiversity science.

Second, many countries, particularly those in tropical regions, understand that their vast inventory of biodiversity is a potential contributor to their long-term economic growth; for example, through ecotourism or bioprospecting for potential new drugs or other products produced by living things. Moreover, these countries also understand the importance of these resources in providing clean water, food, housing materials, and other essential needs of their people. Thus, they can sometimes foster biodiversity science, including systematics, by tying it to these societal needs.

The world's natural history collections contain a huge storehouse of information on Earth's biodiversity. Most specimens have associated information such as where they were collected, their sex, the date they were collected, and so on. The problem is that we have yet to catalog the vast majority of these several billion specimens electronically. If this were done and the information made available to the world's nations, resource management would become much more efficient and cost-effective and move us far toward sustainability. The work of the National Commission for the Conservation and Use of Biodiversity (CONABIO) in Mexico provides a good example of the potential of such information. In recent years, scientists from CONABIO have created their own database, using information on specimens collected from Mexico but contained in large museums around the world. In this way, they have amassed a database of many thousands of records; they can now plot the collecting locality of those specimens on maps that also depict land use in Mexico. This has given them a powerful tool for making decisions about conservation priorities, including where to place reserves, where biodiversity is being lost, and which species are most threatened. If governments of the wealthy countries can accelerate the entry of data from these large collections, it will have a global impact on the management of biodiversity.

Another approach is to build systematics capacity directly. In Bangladesh, one of the poorest countries in Asia, the United Kingdom has constructed a new national herbarium (an institution containing a collection of dried plant specimens); has equipped it with laboratories, a library, and modern equipment and electronic communication systems; and has set up a training program. For its part, the government of Bangladesh is providing operating costs, fourteen scientific staff, and technical and support personnel. This effort is noteworthy because for very little initial investment (only about $2 million), it was possible to create a place for scientific research and training that will benefit Bangladesh well into the future. Perhaps other wealthy countries, and natural history institutions in those countries, will follow this example, and recipient countries will commit themselves to maintain such facilities. If so, this could improve capacity around the world substantially and quickly.

Another way developing countries can build biodiversity science and systematics is to form regional cooperative programs. For example, botanical institutions in eleven southern African nations have formed SABONET, the Southern Africa Botanical Network. SABONET is using funds from the World Bank and the U.S. Agency of International Development (USAID) to build systematics capacity. This capacity includes improved and expanded infrastructure, training, inventorying, databasing, and developing information networks. A major goal of the project is to strengthen the core group of professional and paraprofessional botanists in each country. This group can then undertake programs of inventorying and monitoring, strengthen the management of collections, and expand training. Regional cooperation as exemplified by SABONET makes sense, especially when many countries lack sufficient capacity to undertake the research or training program on their own.

Finally, scientists themselves can initiate international programs that will improve systematics science. A worldwide consortium of systematists, Systematics Agenda 2000 International (SA2000I), has been established within the international biodiversity science program Diversitas. The activities of SA2000I focus on the major research areas of systematic biology already discussed. Systematists involved in SA2000I are organizing regional workshops to help countries establish priorities and needs for biodiversity surveys, seeking ways to expand our understanding of the interrelationships among organisms, and working to improve the storing of biodiversity data in electronic information systems. One successful ongoing effort within SA2000I is Species 2000, an international initiative to assemble a scientifically reliable database of all the world's described living species. Upon entering the names of these species, a user will be able to access other databases housing information about them. This program is very important for managing what we know about biodiversity.

We have discovered only about five percent of Earth's species. This five percent provides us trillions of dollars of economic activity while sustaining and improving our lives. Therefore, it seems very likely that we will gain immeasurably—in new uses and benefits— if we significantly increase our knowledge of the remaining ninety-five percent. Only by improving biodiversity science can we expect to reap the benefits of nature's treasury.

Wilderness Preservation Act, U.S.A.

John Thomas

154 155

The Wilderness Preservation Act, passed in 1964, established the National Wilderness Preservation System and set forth criteria for designating lands for inclusion in this system. According to the act, wilderness must have the following characteristics: absence of significant human impact; opportunities for solitude and primitive recreation; a size of at least 5,000 acres; and outstanding ecological, geological, or scenic value.

Wilderness status protects an area from the building of roads and dams, timber cutting, the operation of motorized vehicles, and new livestock-grazing and mining operations. There are currently about 475 wilderness areas, covering 104 million acres. Although this may seem like a lot, it represents only four percent of the United States land area; moreover, two-thirds of this designated wilderness lies in Alaska.

A recent addition to the wilderness system is an eight-million-acre area under the California Desert Protection Act of 1994, and the next major addition may come in the spectacular redrock canyonlands of Utah. Some of that potential acreage lies within Grand Staircase–Escalante National Monument, which was established in 1996. Congress is debating legislation to give wilderness designation to millions of acres in Utah, but quick resolution is unlikely.

Wilderness is vital to the preservation of biodiversity because these lands protect the genetic variety within species, as well as the ecosystems that they depend on. The United States has more than 260 distinct ecosystem types, and only 160 are currently protected in wilderness areas. But even these protected ecosystems are threatened by degradation from air and water pollution, mining and logging operations, and livestock grazing.

Besides protecting biodiversity, wilderness provides irreplaceable ecological services that human society requires. Watersheds supply us with drinking water, and forests help control global warming by absorbing carbon dioxide from the atmosphere. The wealth of species living in these wilderness areas may even hold the secrets to future medical cures. The study of undisturbed wildlands will also allow us to understand better how human activities have altered the North American landscape. This knowledge is essential if we are to restore degraded lands and protect the wilderness that remains.

How to Grow a Wildland: The Gardenification of Nature

Daniel H. Janzen

Introduction

In 1985, tropical conservation fund-raising centered on the argument that we must buy forest urgently, because cutting it down destroys it forever. Although there is much truth to this, Winnie Hallwachs and I argued that one type of tropical forest—tropical dry forest—could also be restored (and now we know that many kinds of forests can be). Before being largely destroyed by human settlement, tropical dry forest had once covered at least half the forested tropics. The key to restoring a

Daniel H. Janzen is Professor of Biology at the University of Pennsylvania.

dry forest ecosystem was to purchase enough trashed forest remnants somewhere to provide a large area, stop the impacts, and let restoration occur through natural processes. That "somewhere" was northwestern Costa Rica, in particular around the 10,000 hectares of land making up Parque Nacional Santa Rosa. In 1989, the idea became the Area de Conservación Guanacaste (ACG),[1] which is how I will refer to it. To restore this forest, we needed to stop the fires set by people; restore the size of the forest; economically integrate ACG at the local, national, and international scales; and pay the bills for this restoration.

Stopping Fires Set by People

All fires in this tropical dry forest needed to be prevented. They were not natural in this ecosystem—all were set by people. The 88,000 hectares of terrestrial ACG contained at least 50,000 hectares of highly flammable old pastures and brushy fields in 1985. Every time a fire passed through it, more woody vegetation was eliminated and replaced with pasture grasses brought originally from Africa. However, the region had not been sufficiently successful as farms and ranches to have become completely cleared. Without further fires, the forest remnants sprinkled over the partly deforested areas would expand to restore the forest.

Stopping the fires involved straightforward methods: supply trucks, tractors, pumps, radios and walkie-talkies, burned firebreaks, fire lookouts, and lots of brooms. However, the firefighters—a dozen locally hired farmers—also needed to be highly motivated. They had to persuade neighbors not to set the fires in the first place. They had to look for fires twenty-four hours a day during the six-month dry season, and to work without letup, if necessary, to put out any fires. Given full responsibility—and enough budget and flexibility—the firefighters

largely met these goals. When necessary they fought fires on private neighboring land, or called for volunteers to help them when a particular fire got out of hand. Today, the brushy ACG pastures and the interspersed dry forest remnants are essentially fire-free and display at least 40,000 hectares of rapidly regenerating young forest. The seeds arrive by water, wind, birds, bats, rodents, deer, peccaries, monkeys, and carnivores.

In 1977, Santa Rosa park managers removed all the domestic cattle roaming wild, but without beginning a fire-control program. As a result, the African pasture grasses grew two meters high and became fuel for fierce annual fires that destroyed the remaining fragments of forest. The lesson was obvious. When the neighboring land was purchased in the mid-1980s, the young ACG left the cattle on the pastures and even rented browsing rights to as many as 7,000 additional cattle. These biotic mowing machines kept the grass down while the new fire-control program got under way. The newly fire-free pastures filled even more rapidly with woody, shade-producing plants than did those without livestock. However, the cattle cannot be kept on site until reforestation is complete because they damage the streams, rivers, and their associated animals and plants.

The wildland garden is not humanity-free and can never be. Hunters at the end of the Pleistocene Epoch (9,000 B.C.) left their indelible footprints everywhere in the New World by extinguishing our mammoths, ground sloths, horses, and many others. More trampling has occurred every year since. The use of the term wildland indicates that the local species, and their relationships to each other, by and large come about through letting them fight, eat, sprout, grow, move, etc., as they will. We can stop people from setting the forest on fire. However, we cannot shield the forest from all

Figure 2: The same view of tropical dry forest as Figure 1 in the middle of the six-month rainy season, July.

outside influences—for example, global warming, pollution by pesticides, reduction in size, or artificial light. Nor can we counteract the result of the conserved wildland being forever a small island surrounded by the agricultural landscape.

Restoring the Size of the Forest

How big would be big enough for a conserved wildland? At 10,600 hectares in 1985, Parque Nacional Santa Rosa was far too small to survive as a dry forest ecosystem—especially given all the visitors that would come once it became a wildland garden. In particular, it also needed to expand to the wetter east. Much of its biodiversity—insects and birds—seasonally migrated to the rain forests and cloud forests at and across the mountains to the east, and returned to the dry forest in the rainy season.

The ACG expanded until the dry forest was big enough and until it contained its dry-season lifeboats at the eastern end. Its expansion stopped wherever it reached agriculturally profitable land. The expansion incorporated other semiconserved wildland islands. All the private lands between these fragments were purchased from people ranging from squatters to absentee landlords and land speculators, with donations from people willing to pay to see the dry forest survive.

As the ACG increased in area, it eventually became large enough to recover quickly from impact caused by controlled activities of human visitors. Moreover, expansion of the ACG into the more moist eastern rain forests and cloud forests helped to solve problems resulting from the drying and heating that the western ACG dry forests are suffering through global warming.

Socially and Economically Integrating ACG at the Local, National, and International Scales

Social and economic integration of the ACG with the locality, nation, and world occurred in diverse ways. A few examples follow:

- The ACG's biodiversity industry and ecosystem industry helped restore the region's economy, which the death of its cattle industry had severely damaged. Money flowed in, and local people got jobs that provided more security and growth potential than did herding livestock and subsistence farming.

- Since 1987, the ACG Biological Education Program has taught basic biology in the ACG wildland habitats to all fourth-, fifth-, and sixth-grade students, and now high school students, who live near the ACG—more than 2,000 per year.

- Restoring the original forest vegetation throughout the ACG is restoring the watersheds for eleven major rivers that provide water for all local towns and for major farming areas.

- Restoring the original forest vegetation throughout the ACG is removing carbon dioxide (a "greenhouse gas") from the atmosphere.[2] The carbon removed by photosynthesis and stored as plant tissue can even be harvested and parked elsewhere—in buildings, furniture, and even underground deposits—so that it does not return to the atmosphere as carbon dioxide.

- The ACG has been a major factor in many field activities by the Instituto Nacional de Biodiversidad (INBio). This institution has accepted a major responsibility for the Costa Rican national biodiversity inventory, for national bioliteracy, and for computerization of biodiversity management.[3]

- The ACG has been a major stimulus and supporter for the rapidly evolving Sistema Nacional de Areas de Conservación (SINAC). SINAC integrates all of Costa Rica's conserved wildlands into eleven consolidated Conservation Areas. SINAC's wildlands are about twenty-five percent of the country and combine many classical management approaches into one, called "save it by using it without destroying it." Ecotourism is Costa Rica's largest crop. The ecotourist— whether a schoolchild from Peoria or a researcher—is a better kind of cow; the Conservation Areas are the pastures. SINAC was founded to forge a peaceful coexistence between the wildland garden and the urban and agricultural regions. Nothing invites neighbors to encroach on a piece of land more than the impression that it is abandoned. Wildland biodiversity must have high national visibility. The citizenry must see the wildland as a national farm.

- The ACG is developing itself as a site for all kinds of valuable research. What species of plants do the caterpillars of rain forest skipper butterflies eat? Will a pharmaceutical company find its gold in a bottle of frozen baby ticks? How fast does an unburned pasture return to forest? There are more questions than the ACG's hundreds of thousands of species. The ACG is the place to explore them.

- A neighboring orange juice company, Del Oro, is paying the ACG for twenty years of biological control agents, water, consulting, orange peel degradation, and isolation from orange pests. The payment is $480,000 in the form of 1,200 hectares of one of the biologically scarcest habitats in Costa Rica, namely the lowland transition forests between Atlantic rain forest and Pacific dry forest. Del Oro's "green" orange juice is penetrating the Costa Rica market, heading for the European and U.S. markets. This reinforces the contemporary Costa Rican attitude of striving for all of its land use being sustainable.

Paying the Bills

The annual operating budget for Parque Nacional Santa Rosa in the mid-1980s was about $120,000, most of which was spent elsewhere. Today's ACG is ten times as large, costs ten times as much to operate, and generates much income and barter for the region. It meets its costs through a combination of payment for services and interest income from its endowment. Part of this endowment came from international donations for the global environmental services provided by the ACG; the rest came from a "debt for nature swap" in which foreign lenders "forgave" loans to Costa Rica in return for the Costa Rican government spending an equivalent amount of money purchasing land and buildings and setting up the endowment for the ACG. Continued public and private funding is necessary if the ACG is to continue providing its biodiversity and ecosystem services. Recent examples of such public/private partnerships include Yellowstone National Park's new landmark biodiversity prospecting agreement directly with Diversa Corporation in California,[4] INBio's biodiversity prospecting contracts with Merck,[5] and the INBio-Cornell-Bristol-Meyers Squibb ICGB five-year project.

Figure 3: Tropical dry forest in the dry season at the western end of the ACG.

The ACG is ten times as large as the original Santa Rosa; this should, and does, bring massive economies of scale. Why then is the annual budget ten times as large? For one thing, there are costs connected with the ACG's making public its efforts; for example, an Internet Web site is not a freebie. It costs money to persuade a university-graduated Costa Rican biologist to spend a lifetime as a fifth-grade teacher in a remote rain forest town that is just today constructing its first gas station. The tropics have a long-standing reputation for being a source of cheap labor. However, when people move out of the pasture and the bean field, and onto the computer workstation, their wages and costs reach those in the "developed" world.

The development of the ACG, and of many other Costa Rican institutions, has made us all aware that it typically costs more to develop biodiversity and ecosystem preservation programs than underdeveloped host countries are prepared to pay. The developed nations need to continue to provide money and people to help develop wildlands such as the ACG. The product benefits us all.

Conclusions

What general conclusions can we draw from the experience of establishing a wildland garden—the ACG?

- Restoring complex tropical wildlands is primarily a social endeavor; the technical issues are far less challenging.

- In the face of normal human behavior, a large, complex wildland, whatever its origin, can survive only if we look at it in a quite novel way. We cannot afford to view the conserved area as "wild"; this will be the same as saying it is up for grabs.

- A large wildland can be sustained only by bestowing garden status on it, with all the planning, care, investment, and harvest that status implies.

- Any use of the "garden" affects it. However, given time and care as to where the user steps, the garden can restore itself, both in its small pieces and as a whole.

- Planning, care, investment, and harvest within the wildland garden are achieved through detailed understanding of biodiversity and ecosystems. However, simultaneously, we must connect the garden socially and politically to its region, to its country, and to the world.

- The "achievable" is an ever-shifting and negotiated n-dimensional hyperspace produced by the intrinsic traits of a specific wildland interwoven with the mosaic of social energies and agendas brought to bear on it. There is no all-purpose formula other than the pursuit of the goal itself.

Put another way, use it or lose it. And when you use it, something has to restore it.

Figure 4: Tropical rain forest in the dry season at the eastern end of the ACG, at the same time of the year as the photograph in Figure 3, and only forty kilometers away.

1   Visit the ACG Web site, http://www.acguanacaste.ac.cr for more information.

2   See Costa Rica's Carbon Reduction Programs at http://www.unfccc.de

3   For more information, see INBio's Web site at http://www.inbio.ac.cr

4   For more information, visit their Web site at http://www.wfed.org

5   For more information, visit their Web site at http://www.inbio.ac.cr

Pangolin (*Manis* sp.)

Andrew Ramcharan, snakebite healer, Belize.

The Belize Ethnobotany Reserve Project

John Thomas

The extensive rain forests in the small Central American country of Belize are facing growing threats from development and logging activities. In response to these threats, in 1988, the Belize Ethnobotany Project was established with the goals of studying and conserving local knowledge of plant diversity and the plants' traditional medicinal uses.

Among the groups organizing this project were the New York Botanical Garden's Institute of Economic Botany, the Ix Chel Tropical Research Foundation, and the Belize Center for Environmental Studies. This project merges the activities of traditional healers, local farmers, pharmaceutical researchers, and ethnobotanists, who study how cultures understand and use plants for both medical and religious purposes.

As part of the project, local healers banded together to form an association to help conserve and teach traditional knowledge and to preserve Belize's plant diversity. One important result was the designation of a 6,000-acre parcel of lowland forest as a reserve. Plans are being drawn up to help manage this reserve as a place where plants used in traditional medicine can be harvested and conserved.

Another goal of this type of conservation unit, known as an ethno-biomedical forest reserve, is to determine how quickly different medicinal plants grow back after harvest. This would allow healers and reserve managers to determine the levels at which these plants can be sustainably extracted. By learning how to manage plant harvests, it may be possible to protect the biodiversity of forest reserves while providing healers and the Belize population with widely used traditional medicines.

Lesser anteater *(Tamandua tetradactyla)*

On Mt. Mikeno in Congo (near Akeley's grave site).

Penelope Bodry-Sanders: Profile

When Broadway actress-singer Penelope Bodry-Sanders answered an advertisement calling for volunteers at the American Museum of Natural History, she did so only with the certainty that it was time for a change from the stage world of The Threepenny Opera, Jesus Christ Superstar, and the New York Shakespeare Festival: "I have always been interested in anthropology and animal behavior, so it was not just a random chance to volunteer at the Museum."

During her twenty years at the Museum, her experience was varied. She supervised and cared for bird colonies; preserved archival photographs, films, memorabilia, and other materials; curated a special exhibition on the history and accomplishments of the great explorer and collector Carl Akeley; and wrote Akeley's biography. As Education coordinator for Discovery Tours, she designed and developed the educational component of the Museum's travel program.

Inspired by Akeley's vision and concerned about massive habitat destruction in Madagascar, she founded the Lower Primate Conservation Foundation. Dedicated to breeding and raising endangered lemurs in captivity, the foundation has purchased land in Myacca City, Florida and the reserve is now home to a colony of free-ranging lemurs that includes thirteen animals of four different species. Says Bodry-Sanders, "If each of us helps save a piece of the natural world, together we can accomplish a lot."

The Economic Value of Earth's Resources

Graciela Chichilnisky

Introduction

What is the economic value of Earth's resources? The overall question of economic value is a central part of the discipline called market economics. According to market economics, the best way of assigning economic values is through the "marketplace." Economists say that a competitive market exists for a particular product when there are many producers and purchasers of the product and none of them controls its price.

Graciela Chichilnisky is UNESCO Chair of Mathematics and Economics and is Professor of Statistics at Columbia University.

Instead, the price emerges through the interaction between buyers and sellers. (Good examples of what economists would call competitive markets are auctions and the New York Stock Exchange.) The resulting market price simultaneously reflects the costs of producing the product and the preferences shown by consumers when they choose between alternative products. In theory, a market price is set when the supply of the product equals the demand for the product.

Under ideal circumstances—which do not necessarily correspond with the real world—market prices lead to the "most efficient" distribution of the world's (scarce) resources. What economists mean by "most efficient" in this context is that no alternative distribution of these resources would improve the lot of some people without simultaneously making other people worse off. Economic theory and some economic evidence suggest that "free market" pricing works well for typical consumer products such as apples and popcorn. (Economists call products of this type "private goods." Once "consumed" by one person, a private good cannot be "consumed" by anyone else.) However, there is increasing unease today about whether the "free market" does a good job in setting prices for natural resources. Physical and biological scientists question economic wisdom in this regard, and the entire matter has become the subject of a popular debate.

Part of the problem of assigning value to many natural resources is that no organized markets exist for them. This problem is acute for air and water. Markets set the prices of mineral water or bottled oxygen gas; however, no markets or market prices exist for the vast bulk of air and water "consumed" by Earth's humans. Sometimes users pay for water; however, government, not competitive markets, typically sets the price for water. Consequently, the price

does not lead to an "efficient" use of water. Indeed, the price of water is typically so low in the United States that people use it extremely wastefully. A further condition makes the possibility of a market for air even more unlikely than a market for water—people cannot choose to "consume" airs of different qualities or to "consume" differing amounts of air. Economists call goods such as air, water, roads, bridges, etc., "public goods" because, unlike private goods, they are available to everyone in about the same amount. Moreover, one person's "consumption" of a public good does not prevent anyone else from "consuming" the same public good. When it comes to establishing values, standard markets do not work well for public goods. Consequently, economic efficiency is lost. The problem of pricing resources is pervasive. In practice, many scarce and valuable resources are free (have prices of zero). For example, the achievement of cleaner water and air has zero economic value in all systems of economic accounting used today.

Obviously, a system of prices that holds that polluted air is just as valuable as clean air is faulty. Having such faulty prices for crucial resources clearly throws into doubt many traditional ways of assigning a quantitative value to economic progress. For example, we burn fossil fuels to produce industrial output. This output has an economic value (people pay for the output), but clean air does not (people do not pay for the air they breathe). Therefore, burning fossil fuels has an unequivocally positive economic value; it counts as economic progress even as it pollutes the air and increases the atmospheric concentration of greenhouse gases that could cause harmful climate change. A similar situation emerges with the world's forests. The destruction of a forest to extract its wood or to grow agricultural products has an unequivocally positive value and counts as economic progress all over the world. In a world increasingly concerned with the survival of its forests and the deteriorating quality of its air and water, this vision of economic progress defies common sense. It is now under scrutiny.

Markets for environmental assets may never emerge, and, even if they do, they may not act efficiently (in the economic sense defined above). Some economists, myself included, are proposing wider notions of economic value in an attempt to reconcile equity ("economic fairness" among all the world's people) and efficiency; we are also attempting to balance the weight given to the interests of people living today with that given to the interests of people who will be living tomorrow. This essay cannot cover all the issues, important as they are. Instead, it concentrates on discussing basic needs and environmental markets. As an organizing theme, I propose that we must now focus on the choice between two, fundamentally different, patterns of economic growth: resource-intensive economic growth and knowledge-intensive economic growth. One works and the other doesn't. Economic progress is not doing more with more—it is doing more with less.

Before suggesting solutions, however, we should understand the nature of the problem. What is driving our unease? Why is the question of the economic valuation of Earth's resources now timely and controversial? What is the source of the problem? Answering these questions requires a brief review of the situation.

The Global Environment Today
We humans, or our immediate ancestors, have lived on Earth for several million years. Yet only in the past fifty years has human activity reached levels at which it has seriously degraded Earth's environment and decreased its overall biodiversity. On the whole, these harmful ecological effects result from the industrial countries (the North) overconsuming environmental resources, which are overextracted in the developing countries (the South). The North houses less than one third of humankind. However, it consumes most of the world's irreplaceable resources (such as fossil fuels, metals, and minerals). It also consumes most of the renewable resources obtained from fertile land (such as wood, livestock, and cotton).

Mineral fuels provide an extreme case of delivery of a primary, nonrenewable resource from developing countries to industrialized countries. The South sends nearly three-fourths of its exports of mineral fuels (coal, petroleum products, and natural gas) to the North; sixty percent of the North's consumption of mineral fuels comes from the South. Latin America exports mostly resources— about seventy percent of its exports are resources—and Africa does so almost exclusively. The United States alone, with less than five percent of the world's population, consumes an enormous quantity of materials. It has followed a voracious trend that accelerated after World War II. For example, the United States consumes yearly twenty-five percent of all the petroleum extracted worldwide.

The pricing of resources is a crucial aspect of the problem. The world's rapid rate of consumption of fossil fuels derives from the low international prices of petroleum. A similar problem is the overuse of forests to provide wood and wood pulp. The lower the prices, the higher the consumption. Why are the world's resources traded at such low prices? Do market prices fail to convey the true value of Earth's resources? If so, how can we improve this situation?

These questions led me in the mid-1970s to create and develop "basic needs" as a central concept of economic development. This concept was to serve as an empirical measure of economic progress in five continents to complement and sharpen standard economic measures in the areas where these fail. Basic needs are those goods and services that humans need to fit into their societies effectively; examples are food, shelter, education, and health. To a certain extent, these needs vary from one culture to another. I proposed that the satisfaction of the basic needs of the population should be a minimum requirement for economic progress, and I explored the connection between basic needs and sustainable development across the world. Subsequently, my concept of basic needs became a standard aim of development: 150 nations adopted it as an explicit objective in the United Nations Agenda 21, at the 1992 United Nations Conference on Environment and Development in Rio de Janeiro. However, despite the international acceptance for development based on meeting basic needs, the problem persists. The question remains: Why have we reached this pattern of overproduction and overconsumption of the world's resources, beyond the point of sustainability?

The Postwar World: Growth and Trade Based on "Inexpensive" Resources
Today's acute global environmental problems first emerged during the past fifty years. Economic activity has been the driving force, the leading cause, of environmental degradation and the loss of biodiversity. The destruction of biodiversity over the past fifty years is leading to a mass extinction like the one in which all dinosaurs, other than birds, perished. The emissions of greenhouse gases followed a similar pattern. From 1860 to 1950, worldwide consumption of fossil fuels released an

Cethosia cyane, female

estimated 187 billion metric tons of carbon dioxide (an average of about 21 billion metric tons per decade); however, over the past four decades, the rate of emissions was seven times as high, amounting to a total of 559 billion metric tons of carbon dioxide (an average of about 140 billion metric tons per decade).

Today's unease reflects the awareness that the environmental problems we face are new, or at least much more severe than previously. What happened over the past fifty years, and why? Fifty years ago, World War II ended in a victory for the United States and its allies, and the United States became forty percent of the world's economy. After the war, the world community created several important international organizations. These included the United Nations, the International Monetary Fund (IMF), the World Bank, and the General Agreement on Tariffs and Trade (GATT). These organizations worked to establish the vision of economic growth of the leading nation, the United States. This vision involved a very resource-intensive pattern of growth corresponding to a rapidly expanding frontier economy, and to the domination of nature through technological change. After World War II, the world's economies adopted the gross national product (GNP) as a universal measure of economic progress. A country's GNP is the sum of the net value of production of all its goods and services computed at their market prices.

Today, all countries report their economic performance to the UN based on GNP. Yet some fundamental resources without which humans could not survive, such as water and fertile soil, have zero weight in the GNP. There are no organized markets for water and therefore no market prices. Nevertheless, according to World Bank reports, usable water is today one of the scarcest resources in developing countries. Similarly, there is no market, and therefore no market price, for

atmospheric quality or for biomass. In GNP terms, critical resources—such as the whole biomass of the planet, its water bodies, and its atmospheric cover—have zero economic value.

International markets have contributed to the problem of misvaluing resources. Since the end of World War II, the world's economy has grown at a very rapid pace. However, international trade outstripped the overall growth of the world economy by a factor of three. This had important consequences, because most of the misvaluing of resources occurs through international markets. Petroleum is a case in point. In most of the world, petroleum is national property. Economists assign a positive value to its extraction and export, based on the market value of the exported grades of petroleum. However, there is no accounting for the exhaustion of the resource base, the depletion of the asset itself. Destroying a forest to export wood or pulp increases GNP and counts as economic progress. In a world concerned about the preservation of forests and their biodiversity, economics values deforestation and the destruction of biodiversity as unequivocal progress. Why?

After World War II, two major theories of economic growth and trade appeared and gained dominance. The "theory of optimal economic growth" held that a rapid population growth and a corresponding increase in the use of resources are economic ideals. The "theory of international trade based on comparative advantage" held that developing countries should emphasize exports of resources and labor-intensive products, while importing technology and capital-intensive goods. (According to the economic principle of "comparative advantage," trade between two regions will be mutually beneficial if each region specializes in those products in which it is most efficient. This is illustrated by the greatly oversimplified example of a world consisting of two countries A and B. If country A can produce lumber twenty percent as efficiently as country B can but can produce computers only five percent as efficiently as country B can, then there will be a higher total value of global productivity in theory if country A produces only lumber exclusively and country B produces only computers.) Because resources can be obtained at lower prices in the developing countries, the World Bank and the IMF provided strong incentives to developing countries to follow resource-intensive development. As a precondition for loans and other important economic incentives, these organizations recommended that developing countries export more resource-intensive products.

The export patterns we observe in developing countries do not follow the law of comparative advantages (at least do not if resources have suitable prices); nor do they follow any other law of economic efficiency. Nor is the world better off in economic terms when the South specializes in the export of resource-intensive products that damage the environment. To understand why these harmful trading practices occur, we need to take a fresh look at why countries trade resources and at how the international economy assigns prices to these resources. We will see that this trade hinges on differences in property rights between the countries exporting their resources and those receiving these exports.

Property Rights, Industrialization, Prices, and Trade

Many traditional societies have successfully managed their common property resources, such as fisheries and forests, using traditional forms of governance. The term common property refers to ownership that a group shares, rather than individual ownership. An example is Valencia's Tribunal de Las Aguas, a local court in Spain that is 1,000 years old. This court still meets weekly to administer costs and allocate the use of the regions' water network. These traditional systems require a small and stable population, where penalties for antisocial overuse of resources can be administered effectively and, if necessary, across generations. Such traditional systems of resource management tend to break down in the period of industrialization, when outsiders move into the common property area. Lacking firm ties to the area, these outsiders can easily move out to avoid paying any penalties for their overuse of resources. Consequently, what was once well-managed common property changes into unmanaged "open access" resources, which can be had for the taking. A first come, first served system prevails.

When people extract a resource from a resource pool, such as a forest, to which there is open access, the only computed cost is that involved in the actual extraction. Often this is only the cost of the labor and energy it takes to cut and remove the trees. No one considers the cost of replacing the trees to ensure the continuation of the forest. No one computes the costs resulting from the loss of the services that the forest provides to human settlements; these services include providing an ecosystem for biodiversity, shelter, stable climate, and food. Because people undervalue these costs, they overestimate the net benefits from extraction. At each market price for a specific commodity, people extract more under open-access regimes than under private property regimes or under traditional managed systems. Therefore, they overextract the resource, which dwindles and often disappears.

As a result, the country with open property resources offers more of the resource to the international market than is economically efficient. At each market price, the quantity offered is greater in an economic system with open access than in a system with private property. This leads to an apparent comparative advantage (as explained earlier) in the production of environmentally sensitive products, even when there is no real comparative advantage.

This explains why developing countries, which typically have ill-defined property rights for environmental resources, export resource-intensive products even if they have no real comparative advantage in such products. (For example, Mexico exports oil to the United States, but Mexico's oil reserves are small, and they are smaller than those of the United States.) It explains why resource-intensive products such as refined oil, wood, and food are exported at such low prices, prices that are below real costs. The countries with well-defined property rights (for example, the United States) have overconsumed resources, while countries with ill-defined property rights (for example, many developing countries) have overproduced them. As a result, the world economy consumes an inefficiently large quantity of resources, because it takes no account of the costs of the resource overuse. In brief, the process of industrialization itself leads to the inefficient patterns of North–South trade that are at the core of the environmental dilemma today. It leads to international prices for resources that are well below the actual costs to society.

The Economic Value of Earth's Resources

One proposal to correct this problem is to modify the way we account for resources. The idea is to report the costs of using the environment within the national accounts. We generally call this green accounting. The procedure requires, however, that environmental assets have proper prices.

Where will prices for forests, water, biodiversity, and clean air come from? Economists say from free markets. There is much merit and optimism in this premise. Under ideal conditions, market prices lead to economically efficient outcomes that represent the preferences of the population. Water markets are considered currently in California, the Chicago Board of Trade already trades rights to emit the air pollutant sulfur dioxide, and I recently proposed global markets for the right to emit specified quantities of the greenhouse gas carbon dioxide.

However, we cannot trade rights to use air and water and to harvest forests unless we know who owns what. We need property rights on forests, water, air, and biodiversity. Can we carefully parcel out the universe and assign property rights on each of its pieces? This seems a tall order, perhaps too tall for the urgency of the global environmental problem.

Theoretically, assigning property rights for Earth's resources among all its people does not affect overall economic efficiency. Rather it is a problem of equity (economic fairness or economic justice) and therefore, in theory, is more a matter of ethics than of economics. However, in reality, the neat separation of economic efficiency from equity may not work in markets in which people trade environmental assets. Understanding why is important. For markets to function efficiently (in classical economic terms), it must be possible for buyers to choose different quantities of the assets being sold. However, this is not possible for many environmental assets. For example, the concentration of carbon dioxide in Earth's atmosphere is relatively uniform and stable; everyone on Earth breathes the same concentration of carbon dioxide. Likewise, the total biodiversity on Earth is the same for us all. These constraints are physical, not economic or legal. For this reason, economists refer to biodiversity and carbon dioxide concentrations as "public goods" and to their sources as "global commons." (The term commons derives from the village green or village commons, where all citizens had an equal right to graze their sheep.) These are, however, unusual public goods, because they are produced not by governments but by each individual in the economy. (For example, carbon emissions are "produced" privately, by people driving their cars, etc.)

A new discovery of classical economic theory (made by our group at Columbia University with colleagues at Stanford University) is that there can be economically efficient markets for privately produced public goods only when there is a particular equitable (fair) distribution of the total property rights corresponding to these goods. We must properly sort out property rights for environmental markets to achieve efficiency. This requires careful market design and will take some time. In emergencies, taxes or bans on the trading of species that are close to extinction may be necessary; for example, trade in elephant tusks, tiger hearts, and American box turtles.

The Knowledge Revolution
Is it possible to reorient patterns of trade and development without interfering with free trade? To a certain extent the answer is yes. The trade strategies followed by the Asian Tigers (Japan, Korea, Taiwan, Hong Kong, and Singapore) and more recently by the Little Tigers (such as Malaysia) provide good examples. These are very export-oriented countries that moved swiftly away from trade based on traditional comparative advantages—such as labor-intensive and resource-intensive products—to the trade of knowledge-intensive products, such as microprocessors, consumer electronics, communications, financial products, and many other technology-based products. Despite their current financial difficulties, these Asian economies have achieved extraordinary performance over the last twenty years and are way ahead of their African and Latin American

counterparts in all economic indicators. They lead what we call the Knowledge Revolution.

A possible development strategy is to emphasize knowledge-intensive rather than resource-intensive sectors. This economic strategy was introduced formally a few years ago, by our Program on Information and Resources, and received impetus from the empirical evidence of world economic development.

The knowledge-intensive sectors mentioned require human capital and know-how rather than large plants and equipment. Moreover, these sectors are often highly competitive and therefore economically efficient. The computer hardware industry is a good example.

Many developing countries have the skilled labor required for knowledge-intensive sectors. Mexico is currently a producer of such electronic products as microchips and software, and India exports $1.5 billion worth of software a year. Software is very labor-intensive and suits the Indian and Mexican economies because its production does not require large capital outlays. And recently, Barbados's government announced its determination to transform the country into an information age society in less than a generation, based on its excellent educational system.

Conclusions: Information and Resources
Knowledge-intensive growth is successful in economic terms. It drives the most dynamic sectors in the world today. For the purposes of this article, however, its most important aspect is that knowledge-intensive growth does not require intensive use of the environment. It is intrinsically compatible with the continuing health of the global environment. Using information and managing resources may be the most important trends in the world economy

today, and if done wisely, they could lead to economic prosperity that is harmonious with the global environment.

How will all this affect the economic value of Earth's resources? As we change our emphasis away from resource production and exports, the world's available supply of resources that are for sale will decrease. Therefore, the prices of these resources will increase. This means that we will price these resources more accurately, and this is as it should be. By undervaluing Earth's resources, we undervalue ourselves.

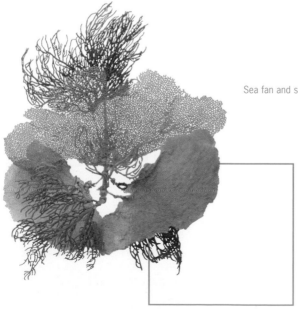

Sea fan and sea rods (*Gorgonia* sp. and *Subgorgia* sp.)

Michael Balick: Profile

For generations, people everywhere have learned about the healing properties of their local plants. As habitat is destroyed and people turn from their old ways, this wisdom may be lost. To help preserve it, ethnobotanist Dr. Michael J. Balick—Vice President for Botanical Science Research and Training and Philecology Curator, Institute of Economic Botany of the New York Botanical Garden; academician; writer; editor; and conservationist—studies how indigenous people use plants around the world. He also works with traditional healers in New York City, observing herbal medical treatments by healers of Chinese and Caribbean origin, and explores the nonmedicinal use of plants: "I am currently working with the Brazilian government and colleagues to domesticate the Babassu palm. It is a palm tree that gives a rich oil, provides charcoal for energy, animal feed, and other products of great value, all of which can contribute to the well-being of the local economy."

Dr. Balick gives the same advice provided by traditional healers in Belize, where he does much of his fieldwork: "It is your world, and you have the responsibility of managing it. Do it from a place of understanding. Be aware of the vulnerability of all species, and the need to maintain a balance in the natural world... Remember, one person really can make a difference."

Glass sponges (*Hyalonema* sp.)

Jaguars Alan Rabinowitz

I never thought about biodiversity in the jungles of Belize, but neither did I take the incredible array of life around me for granted. I awoke every morning before sunrise, listening to the sounds of the night before they were silenced by impending daylight. After breakfast, I left my cabin in the Cockscomb Basin and walked the old timber road that cut a swath through the thick vegetation of the rain forest. My eyes looked toward the ground, searching areas of soft dirt for footprints that would tell me what big animals had passed in the night. In truth, there was only one animal I was interested in, and I would always find its tracks before I was more than 500 feet from my front door. The tracks were unmistakable in their size and shape, and conveyed a feeling of strength and power from the long, even stride they revealed. At this point, the day would start. I was on the trail of my jaguars.

After finishing my Ph.D. in ecology at the University of Tennessee, I came to Belize to conduct a two-month jaguar survey for the Wildlife Conservation Society. By the time I arrived at the Cockscomb Basin, nestled in the Maya Mountains of southern Belize, I was already realizing that Belize contained lots of jaguars. Still, I was not prepared for the abundance of jaguar evidence I initially encountered in the Cockscomb. Nor was I prepared to come face to face with a wild jaguar during the first day's journey into the basin. Strangely, it was not fear I felt on that first meeting with a jaguar, but an almost overwhelming sense of my own smallness compared with the greater biological processes that were going on all around me.

After completing the survey and discussing the results with Archie Carr and George Schaller of the Wildlife Conservation Society, I returned to the Cockscomb to attempt the first detailed ecological study of this species in its rain forest habitat. The decision to return to Belize and study jaguars was not based solely on my feelings for the beauty and majestic nature of this animal, though the memory of that first encounter was never far from my thoughts. However, I had come to realize that, as the biggest carnivore roaming large forested areas of the Neotropics, the jaguar plays an important role in the natural biological processes of the region. The occasional carcasses left by jaguars in the forest or even a cursory examination of their feces along roads and trails clearly indicate that jaguars have an incredibly varied diet. Jaguars will eat monkeys, peccaries, pacas, agoutis, deer, turtles, tapirs, armadillos, iguanas, anteaters, alligators, fish, and almost anything else that crosses their path. Yet monkeys feed on leaves; peccaries feed on roots, seeds, and fruit, and occasionally eat insects; pacas and agoutis feed on fruits; and deer browse on twigs. Their eating habits help control the populations of their food species, just as the jaguars' eating habits help control their populations. Loss of any of the species in this ecosystem, and especially the loss of the highest-level predator—the jaguar—could greatly alter the population sizes of various plant and animal species, possibly even driving some species low on the food chain into extinction. It seemed likely, then, that the jaguars play a key role—perhaps even what ecologists call a keystone role—in the maintenance of the forest structure itself. The forest, in turn, feeds and shelters these animals that the jaguar depends on—and thus feeds and shelters the jaguar itself.

Thus, everything was intertwined. If we focused on protecting a good-sized population of jaguars, wouldn't we then be contributing greatly to the protection of the biodiversity in that population's territory, and thus contributing toward reducing Earth's loss of biodiversity? The jaguars of the Cockscomb Basin were just such a population, but first I needed the scientific data to show people the facts.

Using big box traps, I captured seven jaguars and fitted them with radios around their necks so that I could follow them. With this technique, besides keeping careful notes on all jaguar tracks, scrapes, and feces seen along roads and trails, pieces of the puzzle started coming together. I found out how far jaguars moved, where they slept, and how close they would come to each other. The contents and patterns of their scat droppings (feces) and scrapes (marks in the dirt) told me exactly what they were eating and how they were marking their territories to communicate with each other. I even took a closer look at the abundance of the prey, so that I could see if jaguars were being selective about their food choices. After almost two years, with my research near completion, I realized that there were other more immediate matters to be considered. The Cockscomb Basin was not a protected area, and people outside it were already petitioning the Belizean government for rights to occupy and cut the forest throughout much of the basin.

I had enough data. But now I had to present my findings to the government officials and convince them that protecting a spectacular piece of their natural heritage, and giving jaguars a safe haven, was good for the economy of the country, for the welfare of the people, and ultimately for their own political careers. I had over an hour with the prime minister and his cabinet. I pleaded the jaguars' case as if my own family were being threatened with eviction. I explained that this was not only about jaguars. It was about preserving a piece of the country's biodiversity. If we protected enough forest to support the jaguars, we would be preserving all the plants and animals that lived within the jaguar's range. This, in turn, would ultimately be giving to the people of Belize, not taking away.

At the same time, I took the case to the people themselves. Protection would ultimately mean nothing if it was not accepted by those who had to live and work side-by-side with the jaguars. I talked to cattle ranchers outside the Cockscomb, and explained how healthy jaguars with lots of wild prey will not usually come out of the forest to kill cattle. I talked with the local Mayan communities who hunted and exploited the Cockscomb's resources. I explained that preserving an area that they had taken for granted for so long might restrict some of their activities but would help maintain their plant and animal resources outside the Cockscomb indefinitely.

On December 2, 1984, the minister of natural resources, under instruction from the prime minister, declared the Cockscomb Basin a national forest reserve, with a no-hunting provision for protection of the jaguar. This was the beginning. Belize had just become the first country in the world to protect an area of forest specifically for jaguars. In 1986, the creation of the Cockscomb Basin Wildlife Sanctuary was signed into law, and the area was opened for tourism. Former Mayan hunters who worked with me on my research were now appointed the area's first wildlife wardens. In 1990, and again in 1995, the area under full protection was expanded until it reached nearly 1,000 square kilometers and had been declared "the major achievement in cat conservation for the triennium." By 1995, complaints about jaguars from cattle ranchers in the area had decreased, and the Mayan village at the entrance to the basin had organized village cooperatives to make and sell crafts and medicinal plants to visitors.

Cockscomb Basin today stands as one of the world's success stories in conservation, community participation, and ecotourism. A new generation of jaguars roams the forest relatively unmolested, and the Belizean people speak of the Cockscomb jaguar preserve with pride. My name is rarely mentioned anymore, unless one speaks of the history of the region. And that is how it should be. For the Cockscomb's greatness lies not in the story of any individuals, but in the fact that therein lies a piece of the web of life.

178 179

Wood turtle (*Clemmys insculpta*)

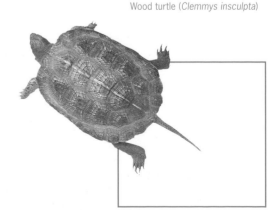

Strategies and Solutions:
Mapping the Biodiversity

Prashant M. Hedao

What is the first thing you do when you plan a trip? You might use a map to find out where the exciting places are and how to get to them. The same holds true about conservation. When you want to do conservation work somewhere that is high in biodiversity, you first need to get a map to show "what is where." Maps are also very powerful graphic tools. Several different layers (types) of data can be presented visually on a single map. It is much easier to understand where something is when you

Prashant M. Hedao is a GIS Software Product Engineer with Environmental Systems Research Institute, Inc.

can see its location on a map than when you have only its geographic coordinates, its latitude and longitude.

Maps have aided decision makers and planners for many centuries—in establishing major townships and cities, building roads and rail networks, land-use planning, and countless other purposes. The history of mapping goes back to the very beginning of civilization. Archaeologists have found records of maps carved on cave walls, stones, animal skins, and parchments. Thanks largely to technological advances (for example, in determining positions on Earth's surface), mapping techniques have greatly improved since ancient times. Today we use maps in more ways than ever before. We use maps for urban land-use zoning; to record the location of such infrastructure components as water mains, sewers, and electrical and telephone cables; to plan transportation networks; to build large dams; to construct electrical power grids; and even to explore for oil and other minerals.

In recent years, mapping has evolved as a powerful tool for conservation planners. It helps them plan conservation activities at different scales—from the continental level to the country level to the very specific local site level. It helps them identify areas of high biodiversity, of species richness, and of endemism (that is, containing species that live nowhere else). It helps them identify any gaps in protection for these areas and to establish protected areas— by looking at the maps of species ranges, habitat blocks, grasslands, wetlands, etc. Conservation planners can overlay maps showing the geographic range occupied by different species with maps showing areas of intact (essentially unchanged) habitat, transportation networks, and socioeconomic data. These multilayered maps can help them

analyze different conservation strategies and choose those most likely to succeed. Maps of this type are typical products of the fast-emerging field called geographic information systems (GIS). In GIS, a computer is used to manipulate and analyze data quickly and accurately and present the results of these analyses in an easy-to-use form.

Apart from the computer hardware and software, data are the most important component of a GIS. The quality of the analysis depends on the quality of the data fed into the system. Researchers can collect these data in many different ways. The most traditional way is to go into the field and record information about species sightings; the field investigator can type this information into the GIS. Once in the computer, these data can be georeferenced (mapped on a world map, country map, etc.) using the latitude and longitude for each data-collection site.

Today, the quickest and most accurate way to learn your geographical position anywhere on Earth's surface is to use the global positioning system (GPS). This consists of twenty-four satellites orbiting Earth. All these satellites contain extremely accurate and precisely synchronized clocks that control the emission of pulsed radio signals; the satellite transmitters continually emit these signals at precisely specified times. On the ground, the signals from whichever four of these satellites are closest are analyzed by a pocket-size, battery-powered computer/radio receiver that has a clock precisely synchronized with the satellites' clocks. By comparing the different times it takes each of these initially synchronized radio signals to reach the receiver, the GPS computer determines its geographic coordinates. The field researcher can then key these coordinates into the GIS system.

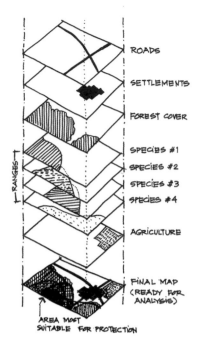

ROADS

SETTLEMENTS

FOREST COVER

SPECIES #1

SPECIES #2

SPECIES #3

SPECIES #4

AGRICULTURE

RANGES

FINAL MAP
(READY FOR
ANALYSIS)

AREA MOST
SUITABLE FOR PROTECTION

GIS layer diagram.

Satellite imagery is another technique that is rapidly emerging as an alternative to the traditional way of surveying the land and making land-use maps and land-cover (type of vegetation) maps. The images and photographs taken by Earth-orbiting satellites provide a good overview of very large geographical areas, an overview that was missing when surveyors used only traditional methods. The technology for interpreting these satellite data is getting simple and easy to use; moreover, extrapolated data are more reliable than they were in the past. Computers analyze the "raw imagery" (the image captured by the satellite's detectors) using different mathematical formulas based on the reflectance index of different types of land cover, and they provide the results in a usable format. (The reflectance index for each type of vegetation shows what fraction of the sunlight striking it reflects off it; the vegetation absorbs the remainder of the incident sunlight. Different types of plants have different reflectivities; we

use their "spectral fingerprints" to distinguish among them.) Researchers should use data obtained from satellite images with great caution; they should always cross-check them with data obtained from ground surveys (also known as ground truthing) for the areas they are less sure about.

The data collected from various sources through the methods cited above should have appropriate "attributes" attached to them to make them "intelligent"; the collection of data is then ready for analysis. Attributes are simply the labels and properties attached to various elements of maps. For example, a road map is more useful—provides more information—as it has more attributes; useful attributes in this case could include the width of the road, the number of lanes, the number of vehicles traveling that section each day, and so on. Similarly, an area of a map that shows the range of a particular species should have an attached attribute specifying that species.

GIS allows us to add attributes to different "layers" of the map. All the layers correspond to the same geographical area, but each layer provides a different type of data. When computers overlay these layers, the attributes corresponding to the same geographical location "interact with each other," and the resulting analysis can thus be powerful. Because a computer can overlay many different relevant layers, GIS is a unique and highly effective tool for planning. Suppose, for example, that we need to identify the right location for a protected area that will include many different species and their habitats—while also avoiding problems caused by intruding humans. This can be done very easily by using GIS and overlapping. The layers we would use here would be the ranges of all species that have been surveyed and range-mapped; remaining habitat data provided by ground-truthed satellite imagery; different transportation network layers such as roads, trails, logging roads, river ferry routes, railroads, etc.; maps of settlements, villages, towns, and cities (with population attributes) in and around the species' ranges; land tenure (ownership, leasing, share-cropping, etc.) and land-use categories of settled regions in and around this area; areas marked for logging; and so on. Overlaying the species-range maps with maps of the remaining habitat tells us what areas are still essentially in their original condition. Furthermore, overlaying the layers of transportation network, human settlement, land use, land tenure, and logging concessions, along with the appropriate attributes, enables us to assess the threat to the habitat blocks.

Based on this type of analysis, we can identify those areas of high biodiversity that are least threatened—and therefore are most likely to survive in their present state—and recommend them as potential protected areas. In other areas that are rich in biodiversity but also highly threatened, conservationists could collaborate with people who have an economic stake in these areas to work out ways to maintain biodiversity by reducing the stress on the environment. ✐

182 183

Fat-tailed Scorpion (*Androctonus australis*)

Community-Based Approaches for Combining Conservation and Development

Nick Salafsky

The Need to Integrate Conservation and Development

Agreement is growing that we face a great crisis: the unprecedented loss of biodiversity. Caused by human activities, this loss is eroding the very resource base on which the survival of all living things, ourselves included, depends. To preserve this resource base, we have to learn how to protect

Nick Salafsky is one of the founders of Foundations of Success.

and manage natural habitats—to use them without destroying them. Biodiversity conservation is thus an essential component of sustainable development.

Simultaneously, we have begun to realize that conservation is first of all a human social issue. Conservation activities are primarily designed to modify human behaviors that affect biodiversity. These threats to the natural world and any potential solutions to the harm they cause are integrally bound up with human development. To address the problem of biodiversity loss, we need to learn how to meet the economic, health, and social needs of an expanding human population. Sustainable development is thus an essential component of biodiversity conservation.

Conservation and development are inextricably linked in a complex system, and maintaining both biodiversity and sustainable economic development poses an immense challenge. To understand how people have attempted to meet this challenge, we need to look at the different approaches to conservation that have evolved over the past century.

The Protected Area Approach

The first formal conservation efforts involved setting up parks and other protected areas that restricted human use of forests and other lands. This practice dates back to ancient times in many cultures from around the world in which wealthy rulers set up private hunting grounds for their own benefit. The current conservation movement, which began in the late 1800s, continued this practice by developing national parks and other protected areas in North America, Europe, and parts of Africa. As shown in Figure 1, the key feature of these protected areas is that they restrict human use of the core biodiversity area. Usually, fences or other obstacles designed to keep people out are set up on the border of the defined protected area. People are supposed to use only resources outside the park, leaving the interior for wildlife; wildlife is supposed to stay inside the park, leaving the surrounding area for people.

Unfortunately, when people began to try this approach in the developing world, they ran into numerous problems. Often, people living around the park were in desperate economic need. No fence or guard patrol could keep them from getting into the park to use its natural resources. Moreover, in many places—for example, in Papua New Guinea and other parts of the Pacific—the local people owned the land and had managed its resources for generations. As a result, it was impractical, immoral, and even impossible suddenly to declare parts of it as off-limits to human use. Finally, in even the most environmentally friendly countries, setting aside more than ten percent of the land as a protected area has not been possible. Besides, the land set aside is often in harsh wasteland that no one else wants—it is no coincidence that the United States has almost no parks in fertile prairie habitats.

Community-Based Approaches to Conservation and Development

In response to these problems, over the past few decades conservationists have worked with communities in developing countries to combine conservation with economic development. Two different approaches have resulted from this collaboration.

Biosphere Reserves

Figure 2 illustrates one approach, which conservationists developed in the mid-1970s.

Paradigm I:
Parks and Protected Areas

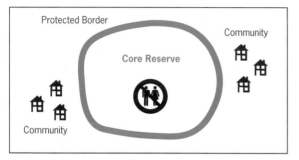

Figure 1: The Protected Area Approach

Paradigm II: Indirectly Linking
Conservation and Community Benefits

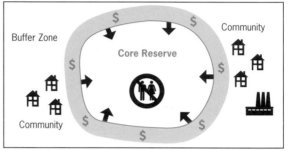

Figure 2: The Biosphere Reserve Approach

Conservation projects using this approach do not restrict their focus to the biodiversity in the center of the reserve. They also try to provide for the economic well-being of the residents surrounding the reserve. For example, the project might help local residents grow coffee in a buffer zone, or it might set up a shoe factory. These broader efforts, combining conservation and economic development over a spatial landscape, became known as biosphere reserves.

Although the biosphere reserve approach addresses some of the problems with the protected area approach, it is far from perfect. In the first place, it does not eliminate the risk of internal threats to the core reserve's biodiversity: local people may continue to hunt or expand their farms inside the reserve while also planting coffee in the buffer zone or working in the shoe factory. In the second place, it creates an economic incentive for expanding the buffer zone into the core area: a farmer

making money growing coffee in the buffer zone will probably expand the coffee garden at the expense of the forest. Finally, the approach does not take into account external threats to the biodiversity. For example, the problem may not be only the local people hunting or expanding their farms; a large logging company might be about to clear-cut the forest, starting at the other side of the reserve. Under the biosphere reserve model, none of the local people may have the personal interest, or the authority, to monitor these threats and take action to protect the biodiversity.

Enterprise-Based Approaches
to Conservation and Development
In the late 1980s, a new strategy for conservation was developed that seeks to address many of the problems of the biosphere reserve approach. As shown in Figure 3, this strategy avoids any attempt to divide the landscape into separate areas for biodiversity or for human activity. Instead, conservationists look at the

**Paradigm III: Directly Linking
Conservation and Community Benefits**

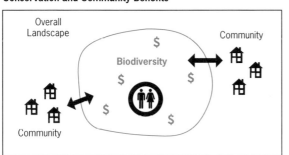

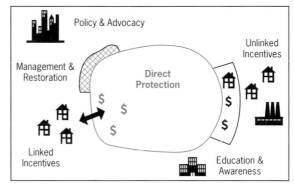

Figure 3: The Enterprise-Based Approach

Figure 4: The Answer—A Mixture of Many Approaches

entire landscape as an integrated whole. The goal here is to try to develop direct links between the biodiversity and the surrounding human populations. The idea is to have humans directly benefit from the biodiversity so they will then have the incentive to identify and take action against both internal and external threats to the biodiversity. For example, the project might help people set up a forest products harvesting business, or it might help them set up a tourist enterprise.

This approach, which is still in the process of being tested, also faces many challenges. It is difficult to set up businesses in remote parts of the world. It is difficult to manage them. And it is even more difficult to ensure that they sustainably use their resource base. Current work indicates that at best, the enterprise-based approach will work only at sites where the major threats to biodiversity are people's need for cash and where local people have the capacity to manage the enterprises.

Lessons Learned

The world is still in the process of learning how to do effective conservation. We clearly do not yet have all the answers. Nonetheless, some basic lessons are emerging from this work.

● Conservation requires more than one approach. As shown in Figure 4, conservation will occur only if we use a mix of different approaches that we tailor to meet the needs of any given site. This includes protected areas, biosphere reserves, and enterprise approaches, along with education efforts, habitat management and restoration, and policy reforms.

● Communities are an important part of the solution, but they are not the solution. Many residents of developed countries have long romanticized the relationship of native peoples in developing countries to their environments. Although local people can indeed be important advocates for conservation, giving local people control of resources does not guarantee that they will conserve them.

- We have to make trade-offs between conservation and development. Although conservation is essential for development and development is essential for conservation, they are not the same thing. Often, development can come only at the expense of conservation, and vice versa.

- The developed countries have much to learn from the experience of developing countries. In the past, conservationists have tried to take lessons learned from the developed world and apply them to the developing world. We are now finding that many lessons can go the other way, too. In particular, we need to think about how to integrate humans and biodiversity in countries such as the United States to supplement the benefit provided by our park system.

- We are still discovering creative ways to make a difference. Combining conservation and development is one of the most difficult problems that people have ever faced. Solving this problem demands the utmost creativity.

If we are going to find solutions to the challenges of combining conservation and development, we cannot just use the latest politically correct "magic bullet." Instead, we need to systematically test a wide range of different approaches and determine the specific conditions under which each will and will not work—and why. We can then use this information to design the appropriate mix of approaches to address each site and each set of problems. Ultimately, finding solutions will require combining all our collective creativity and know-how.

Further Reading

Protected Areas

Kramer, Randall, Carel Van Schaik and Julie Johnson, eds. *Last Stand: Protected Areas and the Defense of Tropical Biodiversity.* Oxford: Oxford University Press, 1997.

Biosphere Reserves

Wells, Michael, and Katrina Brandon. *People and Parks: Linking Protected Area Management with Local Communities.* Washington, D.C.: World Bank, 1992.

Enterprise-Based Approaches

Biodiversity Conservation Network. *Annual Report: Getting Down to Business.* Washington, D.C.: Biodiversity Support Program, 1997.

————. *Evaluating Issues of Business, the Environment, and Local Communities,* 1999. Web site located at www.bcnet.org

General Community-Based Conservation

Margoluis, Richard, and Nick Salafsky. *Measures of Success: Designing, Managing, and Monitoring Conservation and Development Projects.* Washington, D.C.: Island Press, 1998.

Stevens, Stan, ed. *Conservation Through Cultural Survival: Indigenous Peoples and Protected Areas.* Washington, D.C.: Island Press, 1997.

Western, David, and R. Michael Wright, eds. *Natural Connections: Perspectives in Community-Based Conservation.* Washington, D.C.: Island Press, 1994.

In a lower part of the Elwha River still accessible to salmon, some Chinook populations survive.

Restoration of the Elwha River by Dam Removal, Washington

John Thomas

The Elwha River and its many tributaries once flowed uninterrupted for a combined distance of more than seventy-five miles in Washington's Olympic National Park—from the base of 7,000-foot-high peaks to the Strait of San Juan de Fuca, which links Puget Sound to the Pacific Ocean.

In the early 1900s, the construction of two dams blocked the river's free flow. Because neither of these dams had fish passages, ever since then seagoing salmon and trout have been restricted to the lower five miles of the river, now severely degraded as a spawning habitat. Of the ten historical runs of salmon, cutthroat trout, and steelhead trout, the sockeye salmon run has gone extinct. Where fishers once caught 100-pound Chinook salmon, adults now rarely grow to twenty pounds. Overall, the

salmon population has declined from a pre-dam estimate of 400,000 to perhaps 3,000 today.

Yet in 1992, in response to widespread support among the area's fishers, Native Americans, and other citizens, including even the dams' private owners, Congress passed the Elwha River Ecosystem and Fisheries Restoration Act. This historic law requires that the Elwha's salmon and trout populations be restored to their previous abundance. An Environmental Impact Statement assessed different actions, such as removing one or both dams or building fish ladders to allow them to swim around the dams. The analysis determined that the removal of both dams was the only way to restore the populations successfully and enable the entire Elwha River ecosystem to return to its natural state. Plans project that, within ten to twenty years after the Elwha is again a free-flowing river, it can produce 400,000 adult fish annually. The U.S. Department of the Interior's recent proposal to remove these dams—now awaiting final agreement and funding—represents the most significant effort to reverse a century of dam building and to restore the nation's rivers and their native biodiversity.

Large-scale restoration projects such as this do not happen without broad public support. These agreements require difficult legal and political negotiations among government agencies, fishers and recreationists, conservationists, and logging, mining, and industrial interests. Most of the federal funding for dam removal and habitat restoration on the Elwha River has been allocated, but final agreement for release of the funding is still uncertain.

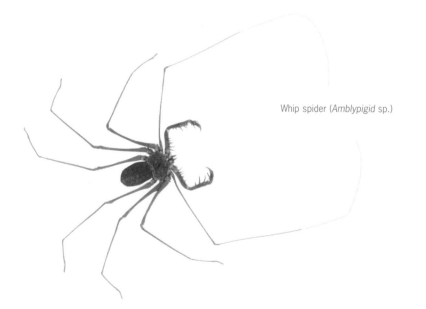

Whip spider (*Amblypigid* sp.)

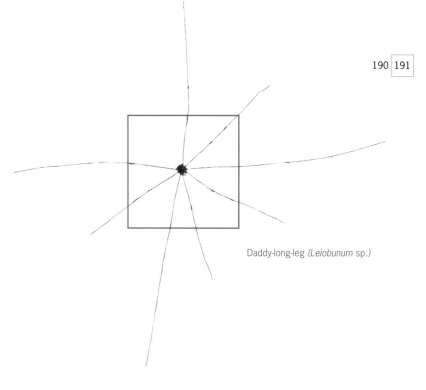

Daddy-long-leg *(Leiobunum* sp.)

No Free Lunch in the Rain Forest

Charles M. Peters

How many times have you heard someone say that she uses
Brazil nut shampoo or copaiba hand lotion, or that he eats rain
forest cookies because using these products preserves the
rain forest since the ingredients are harvested without cutting
down trees? Convincing the American public that products such
as these are harvested in a sustainable manner, and that by
using them we are contributing to the conservation and rational
use of Amazonian rain forests, continues to be a major and
successful marketing strategy.

Charles M. Peters is Kate E. Tode Curator of Botany in the Institute of Economic Botany
of the New York Botanical Garden.

Fruit from *Couma macrocarpa*, a large rain forest tree. The tree is often cut down for its fruit and latex. This method of harvesting is destructive, because it drastically reduces the tree's ability to reproduce.

But although it is true that trees are not being cut down and the forest is not being destroyed, promoting the use of collecting commercial quantities of fruits, oil seeds, and latexes from the rain forest may not be entirely sustainable. In fact, such practices ultimately may harm the forests.

What exactly does "sustainable" mean? From an ecological point of view, a truly sustainable system for extracting nonwood products is one in which rain forest fruit, nuts, oils, gums, latexes, resins, or medicinal compounds can be harvested without reducing the supply of any given species. This means that source species of the nut or fruit or oil must produce a quantity at least equal to that which is being harvested each season.

As is true elsewhere, there are no free lunches in the rain forest. Commercial collectors cannot simply harvest massive quantities of fruits, nuts, and oil seeds year after year and expect the forest magically to replenish these stocks. It is true that picking up fruits and seeds in the forest is a relatively harmless activity compared with harvesting timber, but large, productive trees do not live forever. They need to be replaced with younger individuals on a sustainable and steady basis. In addition, a commercial harvester needs to remember that fruits contain the seeds a tree species needs to reproduce and maintain itself.

If the regeneration and growth requirements of a species are not taken into consideration, intensive, repeated harvesting could cause a valuable resource to become locally extinct. There is already a noticeable lack of seedlings and saplings to ensure a next generation in many stands of Brazil nut trees. This suggests that this species may already have been overexploited.

There are some ecological features of the rain forest that make sustainable harvesting more elusive than it might appear. One of these is the low natural density (number of trees growing close together) and scattered distribution of many tree species. Unlike temperate forests with sometimes vast stands of one or more species, in the rain forest most species are present as two or three individuals per acre. Collectors may have to wander through hundreds of acres in search of large quantities of a particular resource. This harvesting difficulty means that the most accessible and resource-rich patches of forest are usually the first areas to be repeatedly harvested and overexploited.

Another characteristic of many rain forest species is the marked absence of saplings and young trees. Although the canopy may contain several large trees of a particular species, and the understory may be carpeted with hundreds of seedlings, there are few, if any, intermediate-size individuals. New seedlings die naturally for a variety of reasons—lack of light, physical damage, competition from other species, or as meals for plant-eating animals—before they have the chance to become established. But the more seedlings growing on the forest floor, the greater the chance that some will survive to become the next generation. If most of the seeds are collected for commercial use, inevitably there will be fewer, if any, trees that have the chance to grow to maturity.

These two characteristics, low density of individuals and lack of regeneration, make these tree populations difficult to manage and very vulnerable to overexploitation. Collecting fruits and seeds that are likely to die soon after they germinate anyway may seem to have little ecological impact. But how does

Fruit vendor in Iquitos, Peru. *Couma macrocarpa* trunk with gash.

one identify the correct few that need to be left in the forest to maintain the population?

Unlike many temperate-region trees, which rely on wind for pollination and seed dispersal, rain forest trees rely on bats, fish, hummingbirds, beetles, monkeys, bees, and moths to transport their pollen and seeds. The animals depend on the plants for food, and the plants depend on the animals for reproduction.

Pollen, floral nectaries, fruits, and seeds, and the food chains of which they are part form a very complex and poorly understood area of plant-animal relationships. Some of these plant-animal interactions are very specific—one animal may be solely responsible for pollinating a particular type of flower, another may be responsible for the dispersal of a particular type

of fruit. In other cases, "generalist" animals pollinate or disperse seeds of a variety of different tree species.

Given these tree-animal interrelationships, it is very likely that removing large or commercial quantities of fruits, nuts, and oil seeds will have an impact on local animal populations. The problem is that there is no way to predict the nature or magnitude of this impact. Faced with the loss of even some of their food supply, many animals may migrate to more remote areas of the forest, or switch to food sources that are more abundant and available in the same area. Some may be unable to adapt and may even die out. None of these possibilities is good for the exploited tree if the animals that disappear happen to be the ones that pollinate its flowers or disperse its seeds.

Fruit production in rain forest tree populations has a high degree of variability. For example, a single population of fruit trees usually contains a few individuals that produce a few large succulent fruits, a great number of individuals that produce fruits of intermediate size or quality, and some individuals that produce fruits that, from a commercial point of view, are inferior because of small size, bitter taste, or poor appearance.

If these trees are subject to intense fruit collection, those trees with the poor-quality fruit—the "inferior" trees—will likely be the ones whose seeds are left in the forest to regenerate. Over time, the selective removal of fruit from only the best trees will result in a population dominated by trees of minimal economic value.

Last, an especially problematic characteristic of rain forests is that a tree species that has been cut down rarely reproduces itself. Commercial collectors who think that a good way to harvest fruit or latex is to cut down the tree ignore this important fact. For example, leche huayo (*Couma macrocarpa*) is a large rain forest tree that produces commercially viable latex and fruit. Although the species can be lopped as easily as rubber, destructive harvesting in western Amazonia to collect latex has eliminated large numbers of the species. Clearly, no system of extracting nonwood forest products will ever be sustainable as long as harvesting involves an ax.

The commercial extraction of rain forest nontimber products could provide innumerable benefits. But the benefits depend on the forest being able to supply a continuing supply for harvest. If the forest resource is depleted, either through excessive harvesting or the gradual death of economic species with no replacement, no type of new product or new market will make very much difference. Lunch in the rain forest must be paid for, the same way as lunch at the deli. 🖊

194 195

Collecting cama-camu fruit in the Peruvian Amazon.

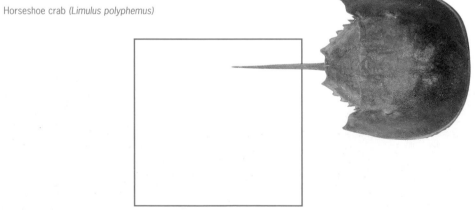

Horseshoe crab *(Limulus polyphemus)*

About the Hall of Biodiversity

The Spectrum of Life wall

On May 30, 1998, the American Museum of Natural History opened a groundbreaking permanent exhibition devoted to what many scientists believe to be the most pressing environmental issue of our time: the need to protect the variety and interdependence of Earth's living things, its biodiversity. The Hall of Biodiversity combines innovative exhibition techniques with new technologies to launch a massive educational effort alerting the public to the biodiversity crisis and its vast implications. The 11,000-square-foot Hall marks an important step forward in the Museum's efforts to expand public understanding of Earth's diverse and often endangered life-forms, and of the critical role this diversity plays in sustaining life as we know it. In addition to informing the public about the biodiversity crisis, the Hall offers a vivid and inspiring vision of the spectacular beauty and abundance of life on Earth.

Every species is interconnected with others in natural systems that shape Earth's atmosphere, climate, and physical features; human beings could not exist without this wide range and variety of life. A survey conducted by the American Museum of Natural History and Louis Harris and Associates before the Hall of Biodiversity opened confirmed that a majority of biologists believe that we are in the midst of a mass extinction of living things—and that expanding human populations and over-exploitation of natural resources are causing an ever-accelerating degradation of enviro-nmental quality and an irreparable loss of species diversity. The survey revealed the scientific consensus that this loss of species will pose a major threat to biodiversity and to human existence in the next century; some scientists say that within the next thirty years as many as half the species on Earth will die in one of the fastest mass extinctions in the planet's 4.5-billion-year history, referred to as the Sixth Extinction.

Scientists at the Museum of Natural History are in the forefront of research on the world's

The Central African Republic Rain Forest diorama

species, working to illuminatc biodiversity through knowledge of the organization, geographic distribution, and history of Earth's species. Such knowledge is essential to stem the tide of extinction and ensure the welfare of the biosphere as governments, scientists, industry, and policy makers throughout the world must take complex collective actions to address species loss. These actions must be informed by accurate scientific information about species diversity.

In their work on biodiversity, Museum scientists draw on one of the world's unique collections of biological specimens. Numbering in the millions, these provide an invaluable record of biological communities and ecosystems over time, and thus document alterations in and responses to environmental stresses. They also contain the primary scientific evidence for the existence and identification of different species, and provide the most reliable documentation of species extinction.

In 1993, in response to heightened international awareness of the importance of biodiversity, the Museum established the interdisciplinary Center for Biodiversity and Conservation. The Center is dedicated to enhancing the use of rigorous scientific data to mitigate critical threats to global biodiversity. To help make the complex political and economic decisions necessary for the survival and protection of global biological resources, the Center provides the essential scientific tools to identify and understand the mechanisms behind the losses, and to equip the world community to use them effectively.

Using a variety of media, ranging from traditional specimens, artifacts, and text to innovative video productions, audio recordings, and interactive computer stations, the Hall defines biodiversity, demonstrates its importance to life as we know it, examines the major issues involved in sustaining it, and showcases the magnificent variety and breathtaking beauty of life.

Video Tour of Earth's Ecosystems

Introduction to the Hall

An eight-minute video introduction to the Hall, narrated by Tom Brokaw, provides visitors with an illuminating overview of biodiversity and the role it plays in the continued balance of the biota.

Biobulletins

By its very nature, the biodiversity issue is continually changing, and cannot be sufficiently addressed by traditional, retrospective museum techniques. In the Hall of Biodiversity, an electronic BioBulletin offers a frequently updated look at the effects on biodiversity of events such as fires, deforestation, global warming, habitat fragmentation, and El Niño. Explanatory commentary accompanies all reported events. This allows the Hall to serve as a resource where the public—on an ongoing basis, on-site and off—can learn about current scientific research and discovery, and about current threats to biodiversity.

The Rain Forest Diorama

The exhibits in the Hall reflect the two "faces" of biodiversity: ecological and evolutionary biodiversity. Ecological biodiversity is illustrated by a 2,500-square-foot diorama that re-creates a portion of the rain forest of the Central African Republic. The diorama includes more than 160 species of flora and fauna and more than 500,000 leaves, each of which was painstakingly detailed by hand. A new form of museum diorama—measuring ninety feet long, twenty-six feet wide, and eighteen feet high—this exhibit joins new technologies, including true-motion video, to depict the interactions of animals and humans with the environment. Through high-resolution imagery, video, sound, and even smell, visitors are immersed in a habitat in which animals appear to move through the forest. Lighting effects simulate the forest ambiance at different times of day. The rain forest is shown in three different states: pristine, altered by natural forces, and degraded by human intervention. In a major departure from tradition, visitors are invited behind the glass wall that usually forms a barrier between viewers and the scene depicted, and thus become a part of this ever-changing and severely threatened ecosystem. Video and text panels alongside the diorama explain many of the details represented within.

The Spectrum of Life

The evolutionary aspect of biodiversity is illustrated by a grand assemblage of more than 1,500 specimens and models, mounted in a 100-foot-long installation along one wall and extending out overhead, vividly conveying the awe-inspiring diversity of life on Earth. Encompassing twenty-eight living groups, resulting from 3.5 billion years of evolution, the creatures in the Spectrum of Life range from microorganisms to terrestrial and aquatic giants, and include bacteria, plants, fish, mammals, and insects. A series of ten interactive computer stations adjacent to the installation identify the specimens featured in the Spectrum, and explain their distribution on Earth. Touch models help visitors better understand such important ecosystem services as erosion, photosynthesis, decomposition, pollination, and other processes.

Video Tour of Earth's Ecosystems

A multi-screen video installation, sixty feet in length, provides a global tour of nine distinct ecosystems, illustrating the breadth and beauty of Earth's habitats while exploring their relative status of peril and preservation. The ecosystems examined in this installation are tundra; islands; temperate and boreal forests; tropical forests; open oceans; coral reefs and coastal wetlands; grasslands and savannas; deserts; and freshwater wetlands, rivers, and lakes. Each

video moves from pristine to degraded habitat; the ecosystems are further explored with texts and maps of their distribution.

The Crisis Zone

Embedded in a case in the floor in the center of the Hall, a timeline of the five previous extinctions includes examples of species lost, while a nearby display case contains examples of endangered and extinct creatures, with information on the causes of their endangerment or extinction. An explanation of biodiversity, clarification of its importance, and a general overview of the causes of the Sixth Extinction are cited on two columns flanking the timeline.

Other Exhibits

The Resource Center presents regularly updated information about how people's everyday decisions affect biodiversity. A bank of ten computer stations offer interactive activities, access to relevant Web sites, a bibliography of information on biodiversity, a database of individual solutions and involved organizations, and a searchable archive of the BioBulletins.

The Transformation Wall consists of text, graphics, and video examining transformations of the biosphere, including those caused by agriculture and urbanization, overexploitation of resources, introduced species, and global environmental change.

The Solutions Wall, designed to be periodically updated, discusses such solutions to the biodiversity crisis as protection and restoration, research and outreach, management of natural resources, and reducing the demand for resources.

Resources

Suggested Reading

Daly, Gretchen C.
Nature's Services: Societal Dependence on Natural Ecosystems.
Washington, D.C.: Island Press, 1997.

A compilation which investigates the relationship between humans and the biosphere. Ecosystem services such as flood control, soil fertility, and food resources are described and discussed.

Dobson, Andrew P.
Conservation and Biodiversity.
New York: Scientific American Library, 1996.

An introduction to the issues associated with biodiversity which concentrates on its scientific and economic value. A lucid handbook which makes biodiversity and conservation accessible to all.

Eldredge, Niles.
Life in the Balance.
Princeton, NJ: Princeton University Press, 1998.

Eldredge takes the reader on a tour of the world's species and ecosystems and explains how we must strike a balance between human needs and the natural world if we are to lead ourselves away from the path of our own extinction.

Grifo, Francesca T., and Joshua Rosenthal, editors.
Biodiversity and Human Health.
Washington, D.C.: Island Press, 1997.

Grifo and Rosenthal bring together leading thinkers on the global environment and biomedicine to explore the human health consequences of the loss of biological diversity. This publication initiates a dialogue among specialists from such fields as public health, biology, epidemiology, botany, ecology, demography, and pharmacology.

Kellert, Stephen.
The Value of Life: Biological Diversity and Human Society.
Washington, D.C.: Island Press, 1996.

A study of how people value nature which draws on twenty years of research. Kellert looks at how factors like ethnicity, gender, age, employment, and where we live determine our values in terms of biodiversity.

Leopold, Aldo.
A Sand County Almanac.
New York: Oxford University Press, 1989.

The father of wildlife ecology, Aldo Leopold has influenced millions of people around the world with this book. The Almanac is Leopold's reflection on his life of observation, and passion about the natural world.

Patent, Dorothy Hinshaw.
Biodiversity.
New York: Clarion Books, 1996.

In this introduction to the subject of biodiversity, Patent details the inter-connectedness of all living things on the planet and explains the importance of saving biodiversity.

Quammen, David.
Song of the Dodo: Island Biogeography in an Age of Extinctions.
New York: Scribner, 1996.

Quammen examines biogeography—the science of the geographic distribution of life on islands—to extrapolate information about extinctions elsewhere.

Takacs, David.
Idea of the Biodiversity: Philosophies of Paradise.
New York: Johns Hopkins University Press, 1996.

Takacs investigates the definitions and philosophies of the scientists who work in the field of biodiversity and looks at how it affects them, both as scientists and as spiritual beings.

Tudge, Colin.
Global Ecology.
London: The Natural History
Museum, 1991.

Tudge examines the role of adaptation, competition and cooperation, chemistry and genetics, behavior, evolution and population dynamics, together with the often disturbing effects that people have on the delicate balance of nature.

Wilson, Edward O.
The Diversity of Life.
New York: W. W. Norton, 1992.

This excellent introduction to biodiversity describes how the world's species became diverse, why the threat to that diversity is greater than anything we have ever known, and what we can do to address this frightening but inspiring challenge.

Wilson, Edward O.
Naturalist.
Washington, D.C.: Island Press, 1994.

The autobiography of one of the leading authorities on biodiversity, *Naturalist* describes E. O. Wilson's growth as a scientist and the evolution of the science he has helped define. An inspiring and beautifully written book.

Videos, Films and Other Media

The Diversity of Life
National Geographic Society
Educational Services
P.O. Box 98109
Washington, D.C. 20560
1-800-368-2728

This twenty-five-minute video examines the diversity of life on Earth. It stresses the importance of preserving endangered species and habitats, and provides potential solutions to the loss of biodiversity.

Encyclopedia of U.S. Endangered Species
ZCI, Inc.
1-800-769-3723

This CD-ROM has reports on U.S. endangered species, with maps, photos, and a glossary. Presentations by the World Wildlife Fund, the Nature Conservancy, Threatened and Endangered Species Information Institute, and ABC News are included.

Green Means
Environmental Media
1008 Paris Avenue
Port Royal, SC 29935
1-800-368-3382
cnvmedia@hargray.com

This video, consisting of a series of short programs, tells the stories of ordinary people who are making positive contributions to the health of our planet.

Life on Earth
Delta Education
P.O. Box 3000
Nashua, NH 03061-3000
1-800-442-5444

This four-hour-long video is a compilation of programs from the David Attenborough series *Life on Earth*. Teachers can cross-reference to find segments which are appropriate to the subject they are teaching.

"Web of Life" Education Kit
WWF Publications
c/o Acorn Naturalists
P.O. Box 2423
Tustin, CA 92781
800-422-8886

A two-hour coproduction by WWF and WQED, Pittsburgh, *Web of Life: Exploring Biodiversity*, is supplemented by a teacher's guide offering extension information and activities.

Web Sites

The Web sites listed below were active as of Fall 2000.

The American Museum of Natural History
http://www.amnh.org

The American Museum of Natural History's site highlights the Museum's activities, research, and educational outreach programs.

The Center for Biodiversity and Conservation at the American Museum of Natural History
http://research.amnh.org/biodiversity/index.html

The Center for Biodiversity and Conservation hosts symposia and publishes resource guides on biodiversity issues. The Web site also includes information about Museum research and research scientists.

The Center for Environmental Research and Conservation
http://www.columbia.edu/cu/cerc

CERC is a consortium of five education and research institutions, including the American Museum of Natural History. The site provides papers and publications, information about research projects, and links to other sites.

Community-Based Environmental Protection United States Environmental Protection Agency
http://www.epa.gov/ecocommunity

CBEP integrates environmental management with human needs, considers long-term ecosystem health, and highlights the positive correlations between economic prosperity and environmental well-being. The site offers case studies, links to other sites, and a unique search engine.

The Garden Club of America
http://www.gcamerica.org

The Garden Club site lists information and applications for the Club's scholarship programs, as well as links to horticultural societies, botanical gardens, and other related sites.

Global Warming
http://www.edf.org/Want2Help/b_gw20steps.html

A Web site created by the Environmental Defense Fund that describes twenty simple steps to reduce global warming.

The National Audubon Society
http://www.audubon.org

The mission of the NAS is to conserve and restore natural ecosystems, focusing on birds and other wildlife for the benefit of humanity and Earth's biological diversity. The site provides information on local chapters, a virtual tour of a swamp, migration routes, and scholarship programs.

National Wildlife Federation
http://www.nwf.org/nwf

The NWF focuses its efforts on five core issue areas: endangered habitat, land stewardship, water quality, wetlands, and sustainable communities. The site offers in-depth information on these issues and explains how to become involved.

The Nature Conservancy
http://www.tnc.org

The Nature Conservancy is dedicated to preserving plants, animals, and natural communities that represent the diversity of life on Earth by protecting the lands and waters they need to survive. This site explains how to contact local chapters, provides information on interning and volunteering, and answers research and homework requests.

The New York Botanical Garden
http://www.nybg.org

The NYBG operates one of the world's most active research programs in systematic and economic botany. The site provides in-depth information on exhibits, educational programs, online publications and resources, and staff research projects.

The Sierra Club
http://www.sierraclub.org

The Sierra Club promotes conservation of the natural environment by influencing public policy decisions: legislative, administrative, legal, and electoral. This site contains news stories, articles, and information about publications and local chapters.

Wildlife Conservation Society
http://www.wcs.org

WCS, headquartered at the Bronx Zoo in New York City, works to save wildlife and wildlands throughout the world. The Web site offers information on its five New York City locations, including workshops, news stories, and science information.

World Wildlife Fund
http://www.wwf.org

Dedicated to protecting the world's wildlife and wildlands, the WWF uses its Web site to explain how to get involved in conservation issues, lists WWF publications, and has information on the organization's global campaigns.

**World Resources Institute—
Biodiversity Resources**
http://www.wri.org/biodiv/biolinks.html

Committed to the idea that a healthy environment and a healthy economy can co-exist, the WRI site links to a wide range of international institutions which work on biodiversity-related issues.

About the American Museum of Natural History

Founded in 1869, the American Museum of Natural History is one of the world's preeminent institutions for scientific research and education.

Today, under the direction of President Ellen V. Futter, the Museum's scientific, education, and exhibition staff are working to discover, interpret, and disseminate knowledge about human cultures, the natural world, and the universe. Prepared for the challenges of twenty-first century society, the Museum is committed not only to making contributions to science, but to improving science education and enhancing science literacy nationwide.

In 1997, the Museum launched the National Center for Science Literacy, Education, and Technology to take the Museum beyond its walls to a national audience. The National Center uses media and technology to connect people of all ages to real scientists and their work. The purpose of the National Center is to take the Museum's vast resources—collections of some 32 million specimens and artifacts, 43 exhibition halls, a staff of more than 200 scientists, and over 130 years of expertise in educational programming—directly to classrooms, libraries, community centers, and homes throughout the country.

Contributors

Mark Bowen is a freelance writer, photographer, and designer of medical instruments. He has a Ph.D. in physics from MIT and has been a rock and ice climber and mountaineer for twenty years. He conceives of science as a form of exploration, and his main interest in writing is to explore the connections between the two. Bowen is writing two books, one about Lonnie Thompson, the other about AMANDA, a neutrino telescope located one mile beneath the surface of the Antarctic ice cap at the South Pole.

Kefyn M. Catley is an Assistant Professor of Science Education at Rutgers, The State University of New Jersey. He is also a Research Associate at the American Museum of Natural History where he studies spider systematics. His educational research revolves around classroom applications of biodiversity science and evolutionary biology. Catley has studied spider biodiversity in temperate South America, Australia, Europe, and throughout North America. He is the author of numerous scientific papers and popular articles.

Graciela Chichilnisky is UNESCO Chair of Mathematics and Economics and is a Professor of Statistics at Columbia University where she has taught since 1980. Since 1994, she has served as the Director of the Program on Information and Resources at Columbia. Chichilnisky is a U.S. citizen and a native of Argentina, where she introduced the concept of "Basic Needs" that was adopted by 150 nations in the UN Agenda 21 as the central element of their strategies for sustainable development. At the 1995 annual meeting of the World Bank, she proposed the global trading of carbon emissions, which became part of the Kyoto Protocol adopted by 166 nations in December 1997, and an International Bank for Environmental Settlements to regulate this.

Theo Colborn, Senior Program Scientist, directs the Wildlife and Contaminants Program at the World Wildlife Fund. In 1985, she received a fellowship from the Office of Technology Assessment of the U.S. Congress, where she worked on human and ecotoxicological issues. She joined the Conservation Foundation in 1987 to provide scientific guidance for the 1990 book *Great Lakes, Great Legacy?* in collaboration with the Institute for Research and Public Policy, Ottawa, Canada. She held a Chair for three years, starting in 1990, with the W. Alton Jones Foundation, and in 1993 was given a three-year Pew Fellows Award. She edited the book *Chemically Induced Alterations in Sexual and Functional Development: The Wildlife/Human Connection,* released in 1992. She co-authored a 1996 book, *Our Stolen Future,* with Dianne Dumanoski and John Peterson Myers, a popular press version of the 1992 book. Colborn has published a number of scientific papers, testified before the U.S. Senate and House of Representatives, lectured extensively, and served in an advisory capacity to state, federal, and international groups concerning the transgenerational effects of toxic chemicals on the developing endocrine, immune, and nervous systems. She was a recipient of the 1999 Norweigian Rachel Carson Prize and the 2000 Blue Planet Prize. Colborn earned a Ph.D. at the University of Wisconsin–Madison in zoology (distributed minors in epidemiology, toxicology, and water chemistry); an M.A. in science at Western State College of Colorado (fresh-water ecology); and a B.S. in pharmacy from Rutgers University, College of Pharmacy.

Joel Cracraft is a member of the Scientific Steering Committee of Diversitas, an international biodiversity science program to promote basic research and training in biodiversity science and conservation. He is also Curator-in-Charge of the Department of

Orinthology at the American Museum of Natural History and was a participating curator of the Museum's new Hall of Biodiversity. He was primarily responsible for the curation of the Central African Republic rain forest diorama, which occupies a large portion of the Hall. Cracraft is using molecular biology methods to investigate the evolution and relationships of groups of birds in Australia and nearby parts of Asia. He and coworkers have also undertaken studies on the conservation genetics of tigers.

Dianne Dumanoski is an author and environmental journalist, whose credentials in the field date back to Earth Day 1970. At *The Boston Globe* she reported on ozone depletion, global warming, loss of biodiversity, the 1992 Earth Summit in Rio de Janeiro, Brazil, and the negotiations on the Montreal Protocol, an international treaty signed in 1987 to phase out the man-made chemicals attacking the ozone layer. She also wrote *One Earth*, a unique monthly environmental column for the *Globe's* Health and Science section, in which she explored cultural, spiritual, and psychological dimensions of the environmental movement as well as innovative ideas such as "green" taxes. Her work has been cited as a model for environmental and science reporting. Besides winning a variety of awards for her reporting, Dumanoski has been a Knight Fellow in Science Journalism at MIT in 1983–84 and a Fellow in 1993 at the University of Colorado's Center for Environmental Journalism. Since leaving *The Boston Globe*, she has been writing, lecturing, teaching, and working on a new book.

David Ehrenfeld is Professor of Biology at Cook College, Rutgers University, where he teaches ecology. He is the author of *The Arrogance of Humanism* and *Beginning Again: People and Nature in the New Millennium*; and was the founding editor of the journal *Conservation Biology*. His book *Swimming Lessons: Keeping Afloat in the Age of Technology*, will be published in 2001. His wife, Joan Ehrenfeld, mentioned in the article, is a plant and ecosystems ecologist and also a Professor at Rutgers.

Paul R. Ehrlich is Bing Professor of Population Studies in the Department of Biological Sciences at Stanford University. He was born in Philadelphia in 1932, received his Ph.D. from the University of Kansas in 1957, and joined the faculty of Stanford University in 1959. He has carried out field, laboratory, and theoretical research on a wide array of problems ranging from the dynamics and genetics of insect populations, studies of the ecological and evolutionary interactions of plants and herbivores, and the behavioral ecology of birds and reef fishes, to experimental studies of the effects of crowding on human beings. His fieldwork has carried him to all continents, from the Arctic and the Antarctic to the tropics, and from high mountains to the ocean floor. Ehrlich is author and coauthor of more than 700 scientific papers and articles in the popular press and over thirty books, including *The Science of Ecology, The Population Explosion, Betrayal of Science and Reason*, and *Human Natures*. He has appeared as a guest on many TV and radio programs and has given hundreds of public lectures in the past thirty years.

Thomas Eisner is the Jacob Gould Schurman Professor of Chemical Ecology at Cornell University and Director of the Cornell Institute for Research in Chemical Ecology (CIRCE). He is a world authority on animal behavior, chemical ecology, and conservation. A member of the National Academy of Sciences and a fellow of the Royal Society, Eisner is the recipient of numerous honors including the National Medal of Science and the Tyler Prize for Environmental Achievement. He is a well-known nature photographer, and has helped make award-winning film documentaries.

Niles Eldredge is a well-known evolutionary biologist and the author of dozens of books for adults and children, students and scientists, and the general public. The subjects he writes about range from trilobites to patterns of extinction, from evolution to biodiversity. His career has been devoted to reconciling evolutionary theory and the fossil record. In recent years, Eldredge has focused on the mass extinctions of the geological past and their implications for understanding the modern biodiversity crisis and future human ecological and evolutionary prospects. As an undergraduate and as a Columbia University doctoral student he did research at the American Museum of Natural History, and upon receiving his Ph.D. in 1969 he was asked to continue working at the American Museum of Natural History. As a Curator in the Division of Paleontology and head of the team of Curators who made the new Hall of Biodiversity, he teaches and does research.

Francesca T. Grifo's interests center on the conservation of biodiversity, including how scientific results are best integrated into conservation projects, policy, and education. She currently oversees projects that demonstrate how this integration is possible. She has focused on intellectual property rights and benefits-sharing issues related to the commercialization of biodiversity, including how these and other issues relevant to scientists are interpreted through the Convention on Biological Diversity. She has worked closely with an array of institutions in Eastern Europe on national-level biodiversity management and planning. Her recent work has examined the relationships between biodiversity and human health. Grifo was formerly the Director of the Center for

Biodiversity and Conservation at the American Museum of Natural History. She now holds an adjunct faculty appointment at Columbia University.

Prashant M. Hedao received his Master's in Landscape Architecture, with a focus in Ecological Planning and GIS, from the Department of Landscape Architecture and Regional Planning at the University of Pennsylvania. Before coming to the United States, Hedao was awarded his Bachelor's degree in Architecture at the University of Bombay in India in 1984. Subsequently, he worked with a Landscape Architect/ Environmental Planner, Satish Khanna. After graduating from the University of Pennsylvania, Hedao worked under Professor John Radke at the University of California at Berkeley in his GIS research laboratory. Over the last two years, Hedao has been using his expertise in the field of GIS and Ecological planning for Biodiversity Conservation at the World Wildlife Fund (WWF) in Washington, D.C. His latest assignment was in Cameroon, where he conducted a GIS and Remote Sensing training workshop for field biologists and managers from parks and protected areas. He is currently working with Environmental Systems Research Institute, Inc. in Redlands, California, regarded as the world's leading provider of GIS software.

Helen F. James, Department of Vertebrate Zoology, National Museum of Natural History, Smithsonian Institution, Washington, D.C., became interested in archaeology as a child growing up in northwest Arkansas, where her imagination was captured by projectile points and other Amerindian artifacts found near her home. After graduating from the University of Arkansas in 1977 with training in archaeology and biological anthropology, James was hired as a Research Assistant at the Smithsonian

Institution to identify fossil birds from the Hawaiian Islands. Before long, various lines of evidence converged to convince her that most of the fossil birds she was studying had become extinct quite recently, because of ecological stresses associated with prehistoric human settlement of the islands. This so fascinated her that she has worked ever since toward understanding what the native Hawaiian avifauna was like before human settlement, and unraveling the causes of prehistoric ecological collapse on islands worldwide. With her husband and research collaborator Storrs Olson, James returns to Hawaii almost yearly to collect fossil birds and other paleoecological evidence. In 1994, she became a graduate student at Oxford University, UK, where she graduated with a Doctoral degree in zoology.

Daniel H. Janzen is a tropical ecologist specializing in animal-plant interactions, biodiversity and biocomplexity, restoration ecology, management of conserved wildlands, and wildland biodiversity development. He has focused on the dry forests of the Área de Conservación Guanacaste (http://www.acguanacaste.ac.cr) in northwestern Costa Rica for the past thirty-four years. Output from this research can be found at http://janzen.sas.upenn.edu. He shares this activity with his tropical ecologist wife, Winnie Hallwachs, who is cooriginator and coexecutor of these research and management activities. Janzen is also Professor of Biology at the University of Pennsylvania.

Simon A. Levin is the George M. Moffett Professor of Biology and was the Founding Director of the Princeton Environmental Institute at Princeton University. He retains an Adjunct Professorship at Cornell University, where previously he was the Charles A. Alexander Professor of Biological Sciences, Chair of the Section of Ecology and Systematics, Director of the Ecosystems Research Center, and Director of the Center for Environmental Research. Levin has also served as President of the Ecological Society of America and of the Society for Mathematical Biology. He also has won the MacArthur Award and Distinguished Service Citation of the Ecological Society of America, among other honors, and is member of the National Academy of Sciences, a Fellow of the American Academy of Arts and Sciences, and a Fellow of the American Association for the Advancement of Science. He was the founding editor of the journal *Ecological Applications*, and has edited numerous journals and book series. Levin is the author of the recent book *Fragile Dominion* and the Editor-in-Chief of the 5-volume "Encyclopedia of Biodiversity."

Ross D. E. MacPhee received his Ph.D. at the University of Alberta in Physical Anthropology in 1977 and was previously Associate Professor of Anatomy at Duke University Medical Center. Since 1988, he is Curator (and from 1993–1999 Chairman) of the Department of Mammalogy at the American Museum of Natural History. He is known for his paleomammalogical research on island extinctions, particularly those which occurred in Madagascar and the West Indies. His recent work concerns how extinctions occur, particularly those in which humans have been implicated. His books and monographs include *Extinctions in Near Time: Causes, Contexts, and Consequences* (editor) and *Primates and Their Relatives in Phylogenetic Perspective* (editor). MacPhee has published more than seventy papers in various journals including the *Journal of Vertebrate Paleontology, American Journal of Physical Anthropology, Nature, Science, American Museum Novitates, Journal of Human Evolution, Palaeontology, International Journal of Primatology,* and *Journal of Archaeological*

Science. He is a Fellow of the American Association for the Advancement of Science and a member of several scientific societies.

Preston A. Marx is a virologist who has worked in AIDS research for the past eighteen years. He has over 140 publications in the field. His work has focused on the natural source of HIV in African monkeys and apes and on animal models for the heterosexual transmission of HIV. Marx's main areas of research are the simian immunodeficiency virus (SIV) models for AIDS pathogenesis and vaccine development. His laboratory uses the SIV/macaque animal model to better understand HIV mucosal transmission, pathogenesis, and to test candidate vaccines. Marx is a Professor of Tropical Medicine and a Core Scientist at Tulane Regional Primate Research Center of the Tulane Health Sciences Center in New Orleans. He is also a Senior Scientist at the Aaron Diamond AIDS Research Center in New York City.

Curt Meine is Director of Conservation Programs with the Wisconsin Academy of Sciences, Arts and Letters. He has served as a consultant to many local, national, and international conservation agencies and organizations, lectured at universities throughout the United States, and taught at the University of Wisconsin–Madison. Meine is author of the biography *Aldo Leopold: His Life and Work* (1988), editor of the collection *Wallace Stegner and the Continental Vision: Essays on Literature, History, and Landscape* (1997), and coeditor of *The Essential Aldo Leopold: Quotations and Commentaries* (1999). He serves on the editorial board of *Conservation Biology and Environmental Ethics* and on the Board of Governors of the Society for Conservation Biology. He is a member of the World Conservation Union's Crane Specialist Group and is a Research Associate with the International Crane Foundation in Baraboo, Wisconsin.

John Peterson Myers is Director of the W. Alton Jones Foundation, a private foundation supporting efforts to protect the global environment and to reduce the likelihood of nuclear warfare. Myers was trained in biology and psychology at Reed College and in zoology at the University of California-Berkeley, where he received a doctorate in 1979. His early research was in the behavioral ecology of migratory birds and in wetland conservation, fields in which he published many scientific articles. In his research career, he held positions at the University of California-Davis's Bodega Marine Laboratory and the Academy of Natural Sciences of Philadelphia, where he was Associate Curator of Systematics until 1987. From 1987 to 1990 he was Senior Vice President for Science at the National Audubon Society, with responsibilities for research programs in wildlife population, ecosystem management, global warming, energy and the environment, and the ecological impacts of deploying biotechnology, among others. Since joining the W. Alton Jones Foundation in 1990, Myers's work has focused on public policy issues underpinning environmental protection, particularly areas associated with maintaining biodiversity, the impact of environmental pollutants on children's health, the effect of fetal contamination on human development, and encouraging the shift from a hydrocarbon to a hydrogen economy.

Norman Myers is an independent scientist and a consultant on environment and development. He is an Honorary Visiting Fellow at Green College, Oxford University, and he holds visiting professorships at the Universities of California, Harvard, Cornell, Michigan, Cape Town, and Utrecht. He is a foreign member of the U.S.

National Academy of Sciences. He is the first British scientist to have won the Volvo Environment Prize and the Sasakawa Environment Prize; and the second British scientist to receive a Pew Fellowship in Conservation and Environment.

Michael J. Novacek a graduate of the University of California–Berkeley, currently serves as Senior Vice President and Provost of Science and as a Curator of Paleontology at the American Museum of Natural History. Novacek conducts broad-based research in systematics and evolution, drawing on evidence from the fossil record and molecular biology. He has also led or participated in field projects in Baja California, the Chilean Andes, Yemen, and since 1990, the Gobi Desert of Mongolia. The Gobi expeditions include the discovery one of the world's richest dinosaur and fossil mammal sites. In recent years he has been active in several biodiversity initiatives, including Systematics Agenda 2000 and the establishment of the Museum's Center for Biodiversity and Conservation. He has also served the profession as a member of the NSF Advisory Board, President of the Society of Systematic Biologists, and on the Board of the American Association for the Advancement of Science.

Fairfield Osborn, 1887–69, was a Trustee of the Wildlife Conservation Society from 1922 to 1969, Secretary from 1935 to 1940, and President from 1940 to 1968. His father, Henry Fairfield Osborn, had been President of the American Museum of Natural History from 1908 to 1933 and President of the Wildlife Conservation Society as well during that time. As President, the younger Osborn was the true leader of the Society, often in collaboration with Chairman of the Executive Committee Laurance Rockefeller. He headed the Society's World's Fair Committee in 1939; opened several

important new exhibits at the Bronx Zoo, including African Plains in 1941 and the Great Apes House in 1950; established the Conservation Foundation and the Society's biological station at Jackson Hole, Wyoming, in 1948; led planning for a new aquarium, which opened in Coney Island in 1957; wrote several seminal books on the global environment, including *Our Plundered Planet* in 1948; and fostered the Society's international conservation program, particularly in Africa.

Charles M. Peters studied forestry at the University of Arkansas and received his Master's and Ph.D. in Plant Ecology from the Yale School of Forestry and Environmental Studies. He has been investigating the ecology, use, and management of tropical forest resources for twenty years. His fieldwork has taken him to two of the largest and least explored tropical regions in the world, lowland Amazonia and the island of Borneo, as well as to the managed forests of Mayan Mexico. Peters is the author of numerous scholarly papers and articles and is currently the Kate E. Tode Curator of Botany in the Institute of Economic Botany of the New York Botanical Garden.

Peter H. Raven, a native Californian with a lifelong interest in natural history, has built on his scientific research to achieve international prominence in the fields of botany and conservation. Following nine years as a member of the Department of Biological Sciences at Stanford University, Raven came to St. Louis in 1971 as Director of the Missouri Botanical Garden and Engelmann Professor of Botany at Washington University. During more than a quarter of a century in these posts, Raven has led the revitalization of the research, educational, and display programs of the Missouri Botanical Garden. The research programs of the Garden now range throughout the world, with particular concentration on

Latin America and Africa as well as important programs in North America, China, and other parts of the world. In addition to his activities in St. Louis, Raven has served for twelve years as Home Secretary of the U.S. National Academy of Sciences, the most prestigious body of scientists in the United States, and as Chair of the National Research Council's Review Committee for the same period. He is now Chair of the NRC's Earth Life Division, President-Elect of the American Association for the Advancement of Science, and a recent winner of the National Medal of Science. Raven's personal-scientific activities now focus mainly on his coeditorship of the *Flora of China,* a joint Chinese-American international project that is leading to a contemporary account of all the plants in China—which, with about 30,000 species, constitute about an eighth of the total found in the world.

Robert C. Repetto is a visiting Professor in the School of Forestry and Environmental Studies at Yale University. He is a member of the EPA's Science Advisory Board and the National Research Council's Board on Sustainable Development. His publications include many research papers on the economics of sustainable development. Repetto received a Ph.D. in Economics from Harvard University, a M.Sc. degree in Mathematical Economics and Econometrics from the London School of Economics, and a B.A. from Harvard College, where he was a member of Phi Beta Kappa.

Scott K. Robinson is a Professor in and Head of the Department of Animal Biology at the University of Illinois at Urbana–Champaign and a Wildlife Ecologist at the Illinois Natural History Survey. He received his Ph.D. in 1984 from Princeton University and his B.A. from Dartmouth College in 1978. In addition to this work on habitat fragmentation in the Midwest,

he works in the tropical forests of Panama and Amazonian Peru. More recently, Robinson has expanded his research to include grasslands and shrublands. A central focus of his research is on the ecology and evolution of avian brood parasitism and its implications for songbird conservation.

Nick Salafsky is currently working with Foundations of Success, a new organization that helps conservation practitioners use adaptive management to learn about different conversation strategies. Previously, he worked for two years as a Program Office for the MacArthur Foundation's Program on Global Security and Sustainability. Beyond that, he worked for five years as Senior Program Officer/Scientist with the Biodiversity Conservation Network (BCN), which was testing enterprise-based approaches to biodiversity conservation across the Asia/Pacific Region. In particular, Salafsky was responsible for helping BCN and its grantees to develop monitoring and evaluation programs that fed into a systematic analytical framework. Before joining BCN, he spent several years conducting interdisciplinary research in West Kalimantan, Indonesia, studying the forest gardens, a locally developed agroforestry system, from economic, ecological, soil science, and agroforestry perspectives. In addition, in the same region of Kalimantan, Salafsky has conducted primatological studies on the red-leaf monkey and has also researched and written a management guide for the endangered freshwater tidal habitats in the Hudson River Estuary in New York State. His background is in the interface of ecology and economics. He received a B.A. in Biological Anthropology/Behavioral Ecology from Harvard University and an M.A. in Resource Economics and an interdisciplinary Ph.D. in Environmental Studies from Duke University.

Edward O. Wilson, Pellegrino University Research Professor at Harvard University, was born in Birmingham, Alabama, in 1929. Growing up in Alabama and northern Florida he received two degrees from the University of Alabama by 1950. Subsequently, he entered the doctoral program at Harvard University, where he remained first as student, then as postdoctoral fellow and professor over the following decades. Wilson's principal interest is in social insects, especially ants, and biodiversity, which he has studied in the field over much of the world. His best-known works include *Theory of Island Biogeography* (1967, with Robert H. MacArthur), *The Insect Societies* (1971), *Sociobiology* (1975), *On Human Nature* (1978), *The Ants* (with Bert Hölldobler, 1990), The *Diversity of Life* (1992), *Naturalist* (1994), and *Consilience* (1998). He was also the editor of *Biodiversity* (1988), which introduced the term.

Credits

Cover

Tree frog, photograph © Joel Cracraft.

Contents

Globe, PhotoDisc.

Cethiosa cyane, female, photo American Museum of Natural History.

Photo of Spectrum of Life wall specimen, courtesy of James Johnsen, American Museum of Natural History.

Foreward

American Museum of Natural History—circa 1900, American Museum of Natural History.

Preface

Tree frog, photograph © Joel Cracraft.

Photos of Spectrum of Life wall specimens, courtesy of James Johnsen, American Museum of Natural History.

Photo of Spectrum of Life wall specimen, courtesy of James Johnsen, American Museum of Natural History.

Section One

Protea, South Africa, photograph © 2000 Joel Cracraft.

Photo of Spectrum of Life wall specimen, courtesy of James Johnsen, American Museum of Natural History.

Sea horse, close-up of head, American Museum of Natural History.

Photo of Spectrum of Life wall specimen, courtesy of James Johnsen, American Museum of Natural History.

Photos of Spectrum of Life wall specimens, courtesy of James Johnsen, American Museum of Natural History.

Collecting fish in southwestern Ethiopia, photo courtesy of Abebe Getahun.

Madagascar Periwinkle flower, © Chris Hellier/Corbis.

Moroccan solitary bee, photo J. G. Rozen.

Photo of Spectrum of Life wall specimen, courtesy of James Johnsen, American Museum of Natural History.

Photo of Spectrum of Life wall specimen, courtesy of James Johnsen, American Museum of Natural History.

Copyright © 1998 by Carl Safina. Reprinted with permission of National Audubon Society, Inc. "The *Audubon* Guide to Seafood" appeared in the May–June 1998 issue of *Audubon* magazine. To subscribe to *Audubon*, call 800-274-4201.

Deforestation for mining in the Amazon, photo Hugh Raffles.

Satellite images, Rondonia, photo NASA ESIP Program and Michigan State University.

Severely deformed Roseate Tern chick, photo Mary LeCroy.

Armadillo, photo John Pontier/Animals Animals.

Dr. Kevin Browngoehl, photo Edward Goldstein.

Euphydryas editha, photo P. R. Ehrlich.

Photo of Spectrum of Life wall specimen, courtesy of James Johnsen, American Museum of Natural History.

"Hot Spots" originally appeared in *Orion*, 195 Main Street, Great Barrington, MA 01230. Reprinted with permission from David Ehrenfeld and *Orion* magazine.

The Soper River flows into Lake Soper, photo Joan Ehrenfeld.

Cotton grass sedge, photo Joan Ehrenfeld.

Inuit children, photo Joan Ehrenfeld.

Pleasant Village Community Garden, photo Green Guerillas.

Demolition in progress, photo Green Guerillas.

Photo of Spectrum of Life wall specimen, courtesy of James Johnsen, American Museum of Natural History.

Carolina parakeet. Illustration by Mark Catesby. From Mark Catesby, *The Natural History of Carolina, Florida, and the Bahama Islands,* London, 1731-1743, Two volumes. Photo of illustration by Jackie Beckett, American Museum of Natural History © 1997.

Figure of major groups of organisms, courtesy of American Museum of Natural History, illustration by Jason Lee.

Passenger pigeon, illustration by Patricia Wynne.

Amy Vedder, photo © Wildlife Conservation Society headquartered at the Bronx Zoo.

Photo of ferns from the Spectrum of Life wall, courtesy of James Johnsen, American Museum of Natural History.

"Chemical Prospecting: The New Natural History" reprinted with permission from *Nature and Human Society.* Copyright © 1999 by the National Academy of Sciences. Courtesy of the National Academy Press, Washington, D.C.

Ants, feeding at a sugary bait, photo Thomas Eisner.

Hemisphaerota cyanea, photo Thomas Eisner.

Hemisphaerota cyanea, portion of a beetle's oily footprint, photos Thomas Eisner.

Slug, onychophoran, and larvae secretions, photos Thomas Eisner.

Dicerandra fructescens, photos Thomas Eisner.

"That They May Survive: Ecuador," illustration by Dolores R. Santoliquido.

Photo of Spectrum of Life wall specimen, courtesy of James Johnsen, American Museum of Natural History.

Section Two

Photo of Spectrum of Life wall specimens, courtesy of James Johnsen, American Museum of Natural History.

Ammonite Fossil (AMNH 45282). Photo by Jackie Beckett, American Museum of Natural History, © 1995.

The Five Big Mass Extinctions, courtesy Edward O. Wilson. Illustration recreated by David Zilkowski.

The island of St. Lucia, photo James Gilardi.

Fledgling parrot, photo James Gilardi.

Clare Flemming, photo Donald A. McFarlane.

From *Natural Change and Human Impact in Madagascar*, edited by Steven M. Goodman and Bruce D. Patterson; "The 40,000-Year Plague: Humans, Hyperdisease, and First-Contact Extinctions" by Ross D.E. MacPhee and Preston A. Marx; copyright © 1997 by the Smithsonian Institution. Used by permission of the publisher.

Woolly mammoth, from *Animals: 1419 Copyright-Free Illustrations of Mammals, Birds, Fish, Insects, etc.* Copyright © 1979, Dover Publications, Inc., New York.

First-contact extinctions map, illustration by Ross D. E. MacPhee.

"Prehistoric Extinctions and Ecological Changes on Oceanic Islands" reprinted from Islands: *Biological Diversity and Ecosystem Function*, edited by Vitousek, P. M., L. L. Loope, and H. Adsersen (*Ecological Studies* 115), with permission from Springer-Verlag New York and Helen F. James.

Ptaiochen pau, photo Julian Hume. Copyright © by the Smithsonian Institution.

"Brown-Eyed, Milk-Giving... and Extinct: Losing Mammals Since A.D. 1500," excerpted from *Natural History*, April 1997.

Xenothrix mcgregori, photo Denis Finnin. Copyright © by the American Museum of Natural History.

Mammals: The Recently Departed, graphics Joyce Pendola.

Illustration recreated by David Zilkowski.

Polluted river, photo by Bob Edwards, SPL/Photo Researchers, Inc.

Increase in global warming, illustration courtesy of NASA. Illlustrated recreated by David Zilkowski.

Photo of Spectrum of Life wall specimen, courtesy of James Johnsen, American Museum of Natural History.

"Thompson's Ice Corps," excerpted from *Natural History*, February 1998.

Solar-powered ice core drilling at 21,500 feet in Bolivia, photo Todd Sowers, Penn State University.

Photo of Spectrum of Life wall specimen, courtesy of James Johnsen, American Museum of Natural History.

"Nest Gains, Nest Losses," excerpted from *Natural History*, July 1996.

Brown-headed cowbird chick in Red-eyed vireo nest, photo © T. Fink & R. Day/VIREO.

Brown-headed cowbird egg in nest of another bird, photo © Ned Smith/VIREO.

Eastern wood-pewee, photo © J. Heidecker/VIREO.

White-eyed vireo, photo © T. Fink & R. Day/VIREO.

Red-bellied woodpecker, photo © W. Greene/VIREO.

"Case Study: Lake Victoria" based on the article "Biomass versus Biodiversity," Stiassny, Melanie L .J., Friends Third Annual Environmental Lecture and Luncheon. New York: American Museum of Natural History, April 13, 1993.

Extinct Lake Victoria cichlids, illustration by J. Green. From George Albert Boulenger, *Zoology of Egypt: Fishes of the Nile*, London, 1907. Photo of illustration by Jackie Beckett, American Museum of Natural History © 1997.

Photo of Spectrum of Life wall specimen, courtesy of James Johnsen, American Museum of Natural History.

"Hormonal Sabotage" From *Our Stolen Future* by Dr. Theo Colborn, Dianne Dumanoski, and Dr. John Peterson Myers. Copyright © 1996 by Theo Colborn, Dianne Dumanoski, and John Peterson Myers. Used by permission of Dutton, a division of Penguin Putnam Inc.

Otter in water, gull on nest, and common mink, American Museum of Natural History.

The DES Paradigm, illustration copyright © Kirk Moldoff.

Photo of Spectrum of Life wall specimen, courtesy of James Johnsen, American Museum of Natural History.

"Case Study: Reefs in Crisis," excerpted from *Natural History*, December 1997–January 1998.

Agincourt Reef, photo Brian E. LaPointe.

Tourists standing on live coral, photo S. J. Lutz.

Photo of Spectrum of Life wall specimen, courtesy of James Johnsen, American Museum of Natural History.

Section Three

Wilson Island, Great Barrier Reef, photo Brian E. LaPointe.

"The New Geologic Force: Man" reprinted from *Our Plundered Planet*, by Fairfield Osborn. Copyright © 1948 by Fairfield Osborn. By permission of Little, Brown and Company.

Fairfield Osborn, photo © Wildlife Conservation Society, headquartered at the Bronx Zoo.

Photo of Spectrum of Life wall specimen, courtesy of James Johnsen, American Museum of Natural History.

Photo of Spectrum of Life wall specimen, courtesy of James Johnsen, American Museum of Natural History.

Jane Goodall, photo Ken Regan/Camera 5.

Photo of Spectrum of Life wall specimen, courtesy of James Johnsen, American Museum of Natural History.

Humpback whale breaching off the Madagascar coast, photo Howard Rosenbaum.

Photo of Spectrum of Life wall specimen, courtesy of James Johnsen, American Museum of Natural History.

Aldo Leopold, photo courtesy of Aldo Leopold Foundation.

Shells on Spectrum of Life wall specimen, courtesy of James Johnsen, American Museum of Natural History.

Jaime A. Pinkham, photo Thomas G. Matney.

Shells on Spectrum of Life wall, photo Denis Finnin. Copyright © by the American Museum of Natural History.

Plants on Spectrum of Life wall, photo Denis Finnin. Copyright © by the American Museum of Natural History.

Provinces Mountains Wilderness, Mojave National Preserve, California, photo © George Wuerthner.

"How to Grow a Wildland: The Gardenification of Nature" reprinted with permission from *Nature and Human Society*. Copyright © 1999 by the National Academy of Sciences. Courtesy of the National Academy Press, Washington, D.C.

Tropical dry forest in the middle of the six-month dry season, March, photo Daniel H. Janzen.

Tropical dry forest in the middle of the six-month rainy season, July, photo Daniel H. Janzen.

Tropical dry forest in the dry season at the western end of the ACG, photo Daniel H. Janzen.

Tropical rain forest in the dry season at the eastern end of the ACG, photo Daniel H. Janzen.

Photo of Spectrum of Life wall specimen, courtesy of James Johnsen, American Museum of Natural History.

Andrew Ramcharan, photo © Michael J. Balick/The New York Botanical Garden.

Photo of Spectrum of Life wall specimen, courtesy of James Johnsen, American Museum of Natural History.

On Mt. Mikeno in Congo (near Akeley's gravesite), photo George Rowbottom.

Wreathed hornbill—close-up of head in profile, photo Bill Meng, © Wildlife Conservation Society, headquartered at the Bronx Zoo.

Cethosia cyane, female, photo American Museum of Natural History.

Photo of Spectrum of Life wall specimen, courtesy of James Johnsen, American Museum of Natural History.

Michael Balick with Hortense Robinson, photo Gregory Shropshire/IX Chel Tropical Research Foundation.

Photo of Spectrum of Life wall specimens, courtesy of James Johnsen, American Museum of Natural History.

Jaguar, photo courtesy of Dr. Alan Rabinowitz, Wildlife Conservation Society.

Photo of Spectrum of Life wall specimen, courtesy of James Johnsen, American Museum of Natural History.

Researchers taking GPS locations to verify interpretation of satellite data, photo Eleanor Sterling, American Museum of Natural History.

GIS layer diagram, illustration © Prashant Hedao 1999.

Photo of Spectrum of Life wall specimen, courtesy of James Johnsen, American Museum of Natural History.

Workshop held at the Tunquini Biological Station, photo courtesy of Meg Domroese, American Museum of Natural History.

The Protected Area Approach, illustration by Nick Salafsky. Illustration recreated by David Zilkowski.

The Biosphere Reserve Approach, illustration by Nick Salafsky. Illustration recreated by David Zilkowski.

The Enterprise-Based Approach, illustration by Nick Salafsky. Nick Salafsky. Illustration recreated by David Zilkowski.

The Answer—A Mixture of Many Approaches, illustration by Nick Salafsky. Illustration recreated by David Zilkowski.

Lower Elwha River, photo courtesy of Friends of the Earth.

Photo of Spectrum of Life wall specimen, courtesy of James Johnsen, American Museum of Natural History.

Photo of Spectrum of Life wall specimen, courtesy of James Johnsen, American Museum of Natural History.

"No Free Lunch in the Rain Forest" reprinted with permission from *Garden*, vol. 14, no. 6, pages 8–13, copyright 1990, The New York Botanical Garden.

Fruit from *Couma macrocarpa*, photo Charles M. Peters.

Fruit vendor in Iquitos, Peru, photo Charles M. Peters.

Couma macrocarpa trunk with gash, photo Charles M. Peters.

Collecting cama-camu fruit in Peruvian Amazon, photo Charles M. Peters.

Photo of Spectrum of Life wall specimen, courtesy of James Johnsen, American Museum of Natural History.

About the Hall of Biodiversity

American Museum of Natural History—circa 1900, American Museum of Natural History.

The Spectrum of Life wall, photo Enrico Ferorelli © American Museum of Natural History.

The Central African Republic Rain Forest diorama, photo Enrico Ferorelli © American Museum of Natural History.

Video Tour of Earth's Ecosystems, photo Enrico Ferorelli © American Museum of Natural History.

About the Museum of Natural History

American Museum of Natural History—circa 1900, American Museum of Natural History.

Glossary

adaptation
Biological characteristic that improves the chance of survival of an animal and its descendants.

aesthetic
Relating to beauty.

allocate
To assign where portions of a resource will go.

archipelago
A chain of islands.

asset
Something one has that is of benefit.

biodiversity
The variety and interdependence of all living things. Biodiversity includes all living organisms, the genetic differences among them, the communities, cultures, habitats and ecosystems in which they evolve and coexist, and the ecological and evolutionary processes that support them.

biological control
The reduction or management of harmful organisms by using other organisms.

biome
Large area dominated by a certain type of plant community, such as desert, temperate grassland, or coniferous forest.

biosphere
The totality of living things on Earth, along with their habitats—the largest ecosystem.

botanical
Relating to the study of plants.

canopy
In a forest, the zone above the ground with the bulk of the tree branches.

chlorofluorocarbons
Synthetic chemicals used in refrigeration, solvents, and Styrofoam manufacture. In the upper atmosphere they break down, releasing chlorine atoms that destroy ozone.

community, biological
All the species living together in an area.

concession, resource
The rights given by a government to a private organization to extract a resource, such as timber or petroleum, for the organization's benefit.

conservation
The management and protection of the natural world.

conservation biology
The scientific discipline concerned with the study and protection of the world's ecosystems and biodiversity.

Cretaceous
The period of geologic time lasting from 144 to 65 million years ago. The end of the Cretaceous Period coincides with the extinction of the dinosaurs.

deforestation
Removal of forests, usually rapidly and over large areas.

demography
The study of population.

development
In economic terms, the creation of industries in an area which allow a higher standard of living for the inhabitants.

dispersal
The movement of organisms to new areas of habitat.

DNA
A complex molecule, found in every cell, that contains the genetic code. Every organism has variations in the code contained in its DNA.

domestication
The development of an interdependent relationship between humans and another species.

ecology
The study of the interactions of living things with each other and their physical environment.

economy of scale
The advantage that a larger producer or consumer enjoys over a smaller one because of costs that do not increase proportionately with size. For instance, a large lawn does not cost ten times as much to care for as a lawn one-tenth as big because it does not require ten times as many lawn mowers.

ecosystem
A community of interacting organisms and their physical environment.

empirical
Based on observation or measurement, rather than on theory.

endangered
In danger of extinction in the foreseeable future.

Endangered Species Act (ESA)
Legislation, passed by Congress in 1973, which protects listed species.

Endangered Species List
List of species protected under the Endangered Species Act.

eon
A major division of the entire 4.5 billion years of geologic time, lasting half a billion years or more.

epizootic
An epidemic in an animal population.

epoch
An interval of geologic time, a subdivision of a geologic period.

equilibrium (pl. **equilibria**)
Balance.

era
An interval of geologic time; a subdivision of a geologic eon.

estuary
A place where freshwater enters the sea (e.g., at river mouth).

eutrophication
The overfertilization of a body of water through the natural or artificial (human-caused) accumulation of nutrients, ultimately leading to excessive growth of algae. The mats of algae may block the sunlight, causing the death and decomposition of shaded algae, which leads to a bacterial population explosion and oxygen depletion.

exotic
A species introduced by humans into a place where it was not previously found. Many exotics thrive in their new environment when freed from their natural enemies, allowing them to displace native species.

exponential
An increasing rate of change. Exponential processes often go from barely noticeable to astronomical in a short time.

extrapolate
To guess at what is not known by using what is known.

family
A group of related genera. The species of a family are similar, such as those of the grass family.

fixing
See nitrogen fixing.

fossil
The remains or traces of an organism turned to stone by geochemical processes.

fossil record
The form, variety, and distribution of fossils in space and time.

gene
An inherited piece of information contained in a cell's DNA. The information from each gene is used to make a unique type of protein, which has a specific function for the cell.

gene flow
The movement of genes among populations of a species. Genes are carried by pollen, spores, seeds, sperm, eggs, and whole organisms.

gene pool
All the genes within a particular population; that is, total genetic variation.

genetic diversity
The number of different genes in a population that perform approximately the same function, and lead to variation between individuals in that population.

genus (pl. **genera**)
A group of related species. The species of a genus are basically the same kind of organism, such as the pines.

glaciation
The formation of glaciers, vast fields of ice and compacted snow. In the Pleistocene Epoch glaciers covered much of North America and Europe.

global warming
An increase in Earth's average temperature.

greenhouse gas
A gas in the atmosphere that reflects back downward the heat radiated from Earth's surface, keeping the atmosphere and Earth's surface warmer than the frigid temperatures of outer space. Important greenhouse gases are water vapor, carbon dioxide, methane, and nitrous oxide.

guild
A group of species that perform the same function within an ecosystem.

gymnosperm
A plant that reproduces by seeds that are formed in cones or other structures that are not flowers—literally "naked seed."

habitat
The environment where an organism can and does live.

Holocene
The geologic epoch extending from the end of the last Ice Age, 10,000 years ago, to the present.

ichthyology
The study of fish.

insectivore
An animal that eats insects.

isotope
A form of a chemical element that is distinguishable by its atomic weight. The atomic weight of the isotope is written in superscript to the left of the chemical symbol, such as ^{14}C.

keystone
An important species whose presence in an ecosystem seem especially critical in maintaining ecosystem processes and regulating populations of other species.

latex
A milky, bitter liquid found in some plants. Some latexes are used for making rubber.

lichen
A plantlike composite consisting of a fungus and an alga. The alga provides food for the lichen through photosynthesis, and the fungus provides protection for the alga, allowing the lichen to grow in places where other organisms could not.

mangrove
A tree that roots offshore on tropical ocean banks. Mangrove stands protect shorelines from erosion and support many other species, making mangroves keystone species. They are often associated with estuaries.

n-dimensional hyperspace
A mathematical concept used to describe a set of associated ranges of measurements. If a subject has a measured trait that exists as a range, such as the heart rate of a bird, that range can be represented as an interval on a number line. If two associated traits are measured, such as heart rate and breathing rate, they can be plotted on a two-dimensional graph where they would form a sort of patch, representing all existing combinations of the two. Three traits could be represented, using some skill, in a three-dimensional plot, and would form a three-dimensional solid, a "space." (n traits, if n is a number more than three, could not be represented together visually in a single plot; they would form a hyperspace.)

Neotropics
The tropical regions of the Americas, the "New World."

nitrogen fixing
The conversion of nitrogen gas from the atmosphere into essential nutrients that are usable by organisms.

overexploitation
Wasteful killing of a species, usually by hunting or poaching, to well beyond the point at which population levels can be sustained.

ozone layer
The layer of the upper atmosphere, approximately twenty-five kilometers in altitude, that contains maximum concentrations of ozone (O_3).

paleontology
The study of fossils.

passerine
A perching bird or songbird.

pathogen
Any disease-producing agent; such as, viruses, bacteria, or other microorganisms.

period, geologic
An interval of geologic time, a subdivision of a geologic era.

pharmacopoeia
A collection of medicines.

phylum
A major grouping of species with similar special characteristics. An example is the arthropods, a phylum that includes insects, spiders, scorpions, millipedes, crabs, and other jointed-leg animals.

Pleistocene
The geologic epoch lasting from 25,000 years ago to 10,000 years ago, coinciding with the last Ice Age.

pollution
Contamination of air, water, or soil by the discharge of harmful, toxic substances.

population
A group of organisms of one species, occupying a defined area and usually isolated from similar groups of the same species.

quantitative
Using numbers, as opposed to verbal descriptions.

Quaternary
The geologic period lasting from 2 million years ago to the present.

radiocarbon (radiometric) dating
An estimate of age in a fossil based on the ratio of the transformed radioactive carbon isotope to the original radioactive carbon isotope. The higher the ratio the older the age of the object. Because the radioactive carbon changes relatively fast, this method is only accurate for ages on the scale of thousands, rather than millions, of years. Dating more ancient objects requires use of radioactive uranium, lead, or other elements.

rain forest
A dense evergreen forest with an annual rainfall of at least 254 cm.; may be tropical (e.g., Amazon), or temperate (e.g., Pacific Northwest).

range
The area naturally occupied by an individual, population, or species.

recycling, nutrient
The movement of chemical elements essential for life from the environment into plants, up the food chain, and back into the environment, where they can be used again.

reintroduction
To place members of a species in their original habitat.

reserve
An area of land set aside for the use or protection of a species and its habitat.

services, ecosystem
Benefits obtained from intact ecosystems, such as freshwater collection, waste breakdown, or oxygen production.

species
A group of individuals, usually identifiable by a set of distinctive features, with a unique evolutionary history. Classically, the members of a species can interbreed only with each other to produce fertile offspring.

Species Survival Plan
Captive-breeding programs administered by the American Zoological Association.

stratosphere
The layer of the atmosphere that extends from approximately ten to fifty kilometers in altitude from the Earth's surface.

stratum (pl. **strata**)
Layer, usually in reference to geology.

subspecies
A population of a species distinguished from other such populations by certain characteristics.

sustainable
Meeting current needs without losing the ability to meet future needs.

systematics
The science of identifying, describing, naming, and classifying groups of organisms and studying their genealogical histories.

taxon (pl. **taxa**)
Any group in the classification of life, such as a family, phylum, or species.

temperate
Having a climate neither extremely hot nor extremely cold; generally found in the zones between 23° and 66° north or south of the equator.

threatened
Populations or species likely to become endangered in the near future.

tropical
Existing between 23° north and 23° south of the equator, generally having a hot climate.

understory
The plants in a forest growing under the trees.

ungulate
A hoofed mammal; for example, horse, rhinoceros, pig, hippopotamus, camel.

U.S. Fish and Wildlife Services
Federal agency that oversees the implementation of the Endangered Species Act.

viable
Capable of living, developing, or germinating.

vulnerable
A species is vulnerable when it satisfies some risk criteria, but not at a level that warrants its identification as endangered.

watershed
The area of land that collects all the rainwater that flows into a given river or stream.

wetland
A permanently moist lowland area such as a swamp or marsh.

Questions

Section One

Biodiversity: Wildlife in Trouble

In what ways can the economic stability of a highly diversified corporation be compared to the ecological stability of an ecosystem with a high degree of biodiversity?

Over forty percent of all prescription medicines used by North Americans are from substances originally extracted from the world's biodiversity. What are some of these medicines and where do they come from?

What's This Biodiversity and What's It Done for Us Today?

How does sustaining biodiversity benefit people?

Scientists recognize that our knowledge of Earth's biodiversity is limited. What implication does this have for our planet?

Deforestation in the Tropics

How does random and unregulated deforestation in the tropics affect the lives of everyone on earth?

In what situations are international policies more effective than national government policies?

Biodiversity and Human Health

How is the genetic diversity provided by wild species important to world food production?

How does ensuring a wealth of biodiversity benefit human health?

Biodiversity: What It Is and Why We Need It

How can people in developed countries adjust their lifestyles to combat overconsumption of resources?

What role does biodiversity play in maintaining planetary life-support systems such as photosynthesis, cellular respiration, and gas exchange?

Hot Spots

Should ecosystems be conserved for reasons other than species diversity?

David and Joan Ehrenfeld were deeply moved by the natural beauty they found in the Soper Valley. Are there natural places that affect you as strongly?

What Have We Lost, What Are We Losing?

Why should we care about biodiversity loss?

How have natural areas in and around your community been changed by the development of housing, business, or roads within the last five years? Have these changes affected local biodiversity?

Chemical Prospecting: The New Natural History

How do the bacteria that live on the feet of *H. cyanea*, the coagulation mechanism displayed by slugs, and the abilities of *D. frutescens* provide good examples of how much there is for naturalists to discover and learn about the natural world?

What value is there in observing the chemicals and chemical processes of living organisms?

Section Two

Evolution, Extinction, and Humanity's Place in Nature

How have humans changed the surface of the Earth, resulting in the current extinction crisis?

What is a "local ecosystem"? To what extent do you live within yours?

The 40,000-Year Plague: Humans, Hyperdisease and First-Contact Extinctions

How can molecular biology be used to test for pathogens?

Are there any modern diseases that can be described as "hyperdiseases"? Why or why not?

Prehistoric Extinctions and Ecological Changes on Oceanic Islands

What evidence shows that islands actually resist extinctions unless humans are present?

Investigate the effects of introduced species on Hawaii, Madagascar, and the Galápagos Islands.

Thompson's Ice Corps

Why did Thompson choose to study ice cores in tropical glaciers?

What natural disasters do scientists believe are attributable to the El Niños of 1982–83 and 1998, and where did they occur?

Nest Gains, Nest Losses

What kinds of behavioral and physical adaptations help songbirds resist predation and parasitism?

What arguments might motivate people to work toward long-term preservation plans for forest habitat?

Hormonal Sabotage

How do synthetic chemicals affect our lives?

Investigate the tragic story of birth defects caused by thalidomide (a drug prescribed for pregnant women in the 1940s and 1950s).

Section Three

The New Geologic Force: Man

How has modern technology changed human lives and affected biodiversity?

Osborn suggests the need for global environmental planning in 1948. What examples of this practice exist in our contemporary world?

Conservation Biology and Wildlife Management in America: A Historical Perspective

How have approaches to conservation biology in the United States changed over the years?

How does Aldo Leopold's description of ecology in 1933 compare with today's use of the word?

Managing the Biosphere: The Essential Role of Systematic Biology

How is the establishment of systematic collections important to the well-being of developing countries?

Apart from economic arguments, what case can be made for the importance of systematics?

How to Grow a Wildland: The Gardenification of Nature

How did increasing the size and biodiversity of the ACG lead to its increased stability?

Why is it imporant to call a "wild" place a garden?

The Economic Value of Earth's Resources

What are the implications of Dr. Chichilnisky's argument that clean air and usable water have zero economic value in today's world?

Why would a baby born in the United States have 100 to 1,000 times more impact on the planet than a baby born in a developing country?

Strategies and Solutions: Mapping the Biodiversity

How can GIS be used in conservation planning?

Research other ways scientists use satellite imagery in their work.

Community-Based Appraches for Combining Conservation and Development

Sustainability has implications for both people and their natural environment. How can conflicting needs be balanced?

Can you find real-life examples of the mixed approach that Salafsky advocates in this essay?

No Free Lunch in the Rain Forest

What complications are involved in harvesting food from the rain forest?

Survey products in your local supermarket that either come from rain forests or have ingredients that come from rain forests. Pick three and try to find out if they're harvested in a sustainable way.